Evaluation of measurement instruments by meta-analysis of multitrait multimethod studies

Frank M. Andrews

Royal Netherlands Academy of Arts and Sciences
P.O. Box 19121, 1000 GC Amsterdam, the Netherlands

Proceedings of the International Colloquium.
Amsterdam, 13–16 February 1989

ISBN 0 444 85724 9

Koninklijke Nederlandse Akademie van Wetenschappen
Verhandelingen, Afd. Letterkunde, Nieuwe Reeks, deel 143

Evaluation of measurement instruments by meta-analysis of multitrait multimethod studies

W.E. Saris and A. van Meurs, *editors*

North-Holland, Amsterdam / Oxford / New York / Tokyo, 1990

Contents

Preface

Survey questions can be formulated in many different ways: the introduction to the question or the question itself can be long or short; the response scale can have 2, 3, 4 or more response categories; the categories can be labeled or not; continuous scales using numbers or lines can be used; a "don't know" category can be explicitly provided or not; the questions can be presented in a battery of similar questions or not, etc. In fact, there are many more variations in the formulation of a single question.

In addition the way in which the data is collected can also be varied: via personal interviews, telephone interviews, mail questionnaires or the more recently introduced computer procedures, e.g. computer-assisted personal interviewing (CAPI), computer-assisted telephone interviewing (CATI) and computer-assisted panel research (CAPAR).

Many studies have shown the effects of these factors on the results obtained for different questions. It is possible that the distributions of the opinions will be different (Schuman and Presser, 1981; Molenaar, 1986; Billiet et al., 1988) which is of course a problem because it is then difficult to know whether the differences are due to the difference in questions or real changes in opinion. This is especially a problem in comparative research.

Other researchers have indicated that qualitative differences in questions can affect the relationships between variables. Andrews (1984) indicated this very nicely using the following equation:

$$r_{ab}{}^{0} = r_{ab}{}^{t} V_a V_b + E_{ab}{}^{c}$$

where a and b are measures
r^0 is the observed correlation
r^t is the true correlation
V is the validity
E^c is the correlated error

Depending on the size of the validity and correlated error, very different true correlations can lead to the same observed correlations. This point is illustrated in table 1.

table 1: Observed and true correlations between two variables: an example.

observed correlation	validity a	validity b	correlated error	true correlation
.40	.60	.70	.00	.950
.40	.93	.95	.36	.045

It is also possible that completely different observed correlations will be obtained while the true correlations are the same. It can also be shown that validity coefficients have a strong effect on the standard error of the regression or correlation of two variables like a and b in the example above. They also have an effect on the standard error of the mean of a variable and on the standard error of the differences of means. The validity of the measures thus has direct consequences for the precision of the studies and therefore on the size of the samples required.

This shows that method variation will not only have considerable biasing effects but also consequences for the precision of the results. Especially in comparative research, it is necessary to be aware of these problems and to cope with them; otherwise comparisons across methods and also across countries cannot be made. Without these considerations there is a risk of obtaining completely wrong conclusions.

In order to take these effects into account information about the effect of the different methods on the size of the different error components is required. Andrews, in his inspiring 1984 publication in Public Opinion Quarterly, has estimated the effects of some of the different factors mentioned above on the quality measures of survey questions. His results are based on a meta-analysis of the quality coefficients obtained for different types of questions in a large series of survey studies where, for each variable, at least three different formulations have been compared using a multitrait-multimethod design. For a detailed presentation of his approach we refer to the reprint of the 1984 article of Andrews in chapter 1 of this volume.

The attraction of this approach is that he could determine on the basis of his results what level of validity could be expected for a specific question. By validity we mean the variance of the observed variable explained by the theoretical variable of interest. For example, if we have the following information:

mean validity = .65
effect of short intro = .03
effect of short question = .02
effect of 5 point scale = .04
effect of a late position in the interview = -.05

then we can predict with an additive model that a question which satisfies the combination of these characteristics will have on average a validity of .69. If such a result is not good enough and we know also that the positioning of a question at the beginning of the interview would improve the validity by .05, then we can move the question to the beginning of the interview and the predicted validity would then be .79.

This kind of information is of course very useful to the survey researcher. It allows a more rational choice out of all the possible variants of the "same question". It also allows correction for measurement error without repeated observations in each survey.

Although the study of Andrews provides a lot of information about the effects of the different factors, there are at least two reasons why his research should at least be partially replicated.The first reason is that since 1979, when he did his studies, a number of new developments have occurred, for example:
- the use of continuous scales in different forms (Lodge, 1981; Saris, 1988);
- the use of computer-assisted data collection procedures such as CAPI (Sikkel, 1988) and CAPAR (de Pijper and Saris, 1986).

The second reason is that it would be useful to carry out a much larger study with many more factors crossed with each other to get better information about the interactions between them. One interesting factor would be "nation". It would be very useful in comparative research to know whether it matters very much if a question is asked in the USA, Holland, Sweden or Germany.

An international study

Given the high costs of such a study, an international project was organized to make a study along the lines developed by Andrews. In such a joint project we could share the costs, increase the number of factors and considerably increase the generalizability of the results. In this project the research groups of each participating country would do their own research and be free to publish their results but the factors varied in these studies would be determined by the international research group. At the end of the three-year project, all analyses will be combined and an analysis would be done across countries for detecting the effects of the different factors and the interactions between them.

Fortunately, the organizational structure of such a study would be rather simple and would allow each research team a lot of freedom while the analysis across countries could add a lot of extra information to our existing methodological knowledge.

Such a study would also provide the possibility of doing an elaborate study on questions which are very important for comparative research such as quality of life measures or product evaluations. Verbal responses would perhaps be difficult to compare across countries, continuous line and number scales might provide better tools. If such questions were to be incorporated by the groups in all participating countries we would be able to compare validity aspects of different question formulations across countries as well as the substantive results and the combination of these two aspects.

The first meeting of the participants of the international project was held in February 1989 in Amsterdam with the valuable support of the Royal Netherlands Academy of Arts and Sciences. During this conference the different aspects of this project were discussed:

- the model to be used for the evaluation of measurement instruments
- the design of MTMM studies
- the design of the meta-analysis of the MTMM results

As can be seen in the table of contents these topics are the subject of this book. This collection of papers forms the proceedings of the conference. In addition, summaries of the discussions which took place during the conference are presented.

ACKNOWLEDGEMENTS

We would like to thank the Royal Dutch Academy of Arts and Sciences for subsidizing our meeting and Dr. Kwantes and Mrs. A. Muller for all the help they have given us in making it a success. We very much appreciate their efforts. This colloquium was certainly an important event in the design of this international research project which would have been impossible without this support.

During the preparation of this book the help of Annette Scherpenzeel was much appreciated.

REFERENCES

Andrews, F.M. (1984). Construct validity and error components of survey measures: A structural modelling approach. *Public Opinion Quarterly, 48*, 409-442.

Billiet, J., Loosveldt, G., & Waterplas, L. (1988). *Response effecten bij survey-vragen in het Nederlands taalgebied*. Rapport 1988/6. Leuven: Sociologisch Onderzoeks Instituut.

Lodge, M. (1981). Magnitude scaling, quantative measurement of opinions. London, Beverly Hills: Sage Publications.

Molenaar, N.J. (1986). *Formuleringseffecten in survey-interviews (Wording effects in survey-interviews)*. Amsterdam: VU-Uitgeverij.

Pijper, W.M., & Saris, W.E. (1986). Computer assisted interviewing using home computers. *European research, 14*.

Saris, W.E. (1988). *Variation in response functions: A source of measurement error in attitude research*. Amsterdam: Sociometric Research Foundation.

Schuman, H., & Presser, S. (1981). *Questions and answers in attitude surveys*. London: Academic press.

Sikkel, B. (1988). *Quality aspects of statistical data collection*. Amsterdam: Sociometric Research Foundation.

Amsterdam, February 1990

W.E. Saris and A. van Meurs

Part 1

Evaluation of measurement instruments

chapter 1

Construct validity and error components of survey measures: a structural modeling approach*

F.M. ANDREWS

Abstract

Measurement errors can have profound effects on statistical relationships, and better information on the quality of measures seems needed. This study uses a new technology (structural modeling of data from special supplements to regular surveys) to generate estimates of construct validity, method effects (a major source of correlated error), and residual error (mainly random error) for a broad set of measures obtained from five national surveys and an organizational survey (total respondents = 7.706). Analysis of these estimates suggested that a typical survey item, when administered by a respected survey organization to a general population sample, can be expected to yield 50-83 percent valid variance, 0-7 percent method effects variance, and 14-48 percent residual variance. Multivariate analysis showed that over two-thirds of the variation in measurement quality could be explained by 13 survey design characteristics; characteristics of respondents explained a small additional portion.

Results provide:

a. information on design conditions associated with better (or worse) measurement quality
b. empirically based suggestions for improving measurement quality in future surveys
c. a set of coefficients for predicting the quality of measures not studied here.

* This paper is a reprint of the original paper printed in Public Opinion Quarterly (1984) pp. 409-442. POQ kindly permitted this reprint.

Gerald A. Cole and Mary Grace Moore made numerous and substantial contributions to the work reported here. The author is grateful to David Bowers, Angus Campbell, Charles Cannell, Philip Converse, Richard Curtin, Daniel Denison and Robert Groves for allowing us to include methodological supplements in some of their surveys. This research was supported by a grant from the National Science Foundation.

CHAPTER 1

Introduction

There is growing recognition that measurement errors in any kind of data (including survey research data) can have profound effects on statistical relationships. Some kinds of measurement errors make bivariate relationships appear stronger than they really are; others make bivariate relationships appear too weak. The effects of measurement errors on multivariate relationships can be great, and also complex. Under certain combinations of errors, an observed relationship can be "wrong" in both direction and magnitude. However, if one has information about the validity and error composition of the measures being analyzed, more informed judgments can be made about the underlying relationships that are of primary interest.

Insightful survey researchers have always been interested in the quality of their data, and new information about data quality has been a major contributor to the development of survey technology. Much attention has been devoted to sampling errors, and there now exist good ways to estimate their magnitudes and much knowledge about how to reduce them. One kind of measurement error, bias (a consistent tendency for a measure to be higher or lower than it "should be") has also received considerable attention (Sudman and Bradburn, 1974). However, while bias can produce serious distortions in percentages, means, and other measures of central tendency, and hence is a threat that must always be considered, a bias that is constant for all respondents does not affect linear relationships at either the bivariate or multivariate level. It is other kinds of measurement errors that intrude on relationships (random and correlated measurement errors).[1] (Key terms are defined below). These are the kinds of measurement errors investigated in this study.

For each of the measures included in any particular analysis, one would like, ideally, to be able to apportion the total variance into three components: valid variance, correlated error variance, and random error variance. From this, one could know the extent to which the true bivariate relationships (i.e., the relationships among the concepts being investigated) were being attenuated (because of random measurement error) and/or inflated (because of correlated measurement error). In addition, one could sort out the complex effects that random and correlated measurement errors have on multivariate statistics such as regression coefficients,

[1] The conceptualization of measurement quality used here is similar to that discussed by Heise and Bohrnstedt (1970) and by Zeller and Carmines (1980).

multiple and partial correlations coefficients, and path coefficients.[2] A pair of examples, taken from the data of this study, will illustrate how misleading even a simple bivariate relationship between observed measures can be when allowance is not made for the effects of measurement errors. In survey 2, the observed product-moment correlation between items having to do with perceptions about changes in business conditions over the past year and in the coming year averaged .41.[3] After allowing for measurement error, however, the true relationship between respondents' perceptions was estimated to be .70. Thus in terms of overlapped variance, the observed relationship was only about one-third of what it should have been (17 percent versus 49 percent). In this case, random errors led to a gross deflation of the relationship. However, this does not always occur. Survey 5 produced a relationship of .44 between evaluations of own health and of work that had to be done around the house; but after allowing for measurement error, the true relationship was estimated to be .30. Here, correlated error overwhelmed random error, and the observed percentages of overlapped variance was more than double what it should have been (19 percent versus 9 percent).

This study has four major goals, none of which has been pursued previously in a large-scale and systematic way:

1. Test the feasibility of incorporating a particular kind of methodological supplement in regular on-going national and organizational surveys and of using structural model estimation techniques to generate estimates of measurement quality.
2. Provide descriptive information about estimated construct validity, method effects, and residual error for a broad range of survey measures as implemented by the standard data collection procedures of a respected survey organization.
3. Account for why some survey measures have higher (or lower) measurement quality than others.
4. Provide a means for predicting the construct validity and error components of other survey measures not actually examined in this study. These goals lead to a more general outcome of considerable importance to survey researchers and other users of survey data: more knowledge about how to produce better data.

[2] The important impact that measurement errors have on statistics of relationships has received some attention in recent years (e.g., in sociology by Bohrnstedt and Carter, 1971; in psychology by Linn and Werts, 1973; in political science by Asher, 1974; in statistics by Cochran, 1970), but it still goes unrecognized by many data analysts.

[3] This is an average of nine different correlations (ranging from .24 to .55) that assessed this relationship.

CHAPTER 1

BASIC NOTIONS ABOUT VALIDITY AND MEASUREMENT ERROR

Definitions

Validity refers explicitly to construct validity: the extent to which an observed measure reflects the underlying theoretical construct that the investigator has intended to measure (Cronbach and Meehl, 1955; American Psychological Association, 1974).[4] As noted by Zeller and Carmines, construct validity is different from several other types of validity (content, concurrent, predictive) and involves different notions from those of reliability, but "is the most appropriate and generally applicable type of validity used to assess measures in the social sciences" (Zeller and Carmines, 1980 pp. 83).

The difficulty in estimating construct validity arises from its explicit linkage to an unmeasured theoretical construct, and many discussions of construct validity stress the importance of a theoretical model in the construct validation process. Recently developed structural modeling techniques allow theoretical models to enter the analysis in much more explicit and powerful ways than could be achieved before and (as will be described in this paper) such models play a fundamental role in the present study.

Random measurement error refers to deviations (from the true or valid scores) on one measure that are statistically unrelated to deviations in any other measure being analyzed concurrently. Conversely, correlated measurement error refers to deviations from true scores on one measure that do relate to deviations in another measure being concurrently analyzed. Note that wether an error gets classified as correlated or random depends on what other measures happen to be included in a given analysis.[5]

[4] As is common (but not universal) practice, in this study a measure's validity is expressed as the correlation between the measure and the underlying construct; hence the square of this figure will indicate the proportion of valid variance in the measure. (For example, a measure having a validity of .8 would consist of 64 percent valid variance.)

[5] The deviations that constitute random or correlated errors each average to zero across the respondents being examined. If deviations do not average to zero, it is possible to apportion these deviations into two components: a constant, which is the "bias" discussed previously-an important source of error in its own right, but which does not affect realtionships and is not adressed in this investigation- and the remaining part, in which deviations do average to zero across respondents. This use of the term correlated error emphasizes an important general aspect of these errors. Other phrases sometimes used to refer to the same phenomenon include systematic error and halo effects. However, our use of correlated error is different from the way the term has sometimes been employed in investigations of interviewereffects-i.e., the extent to which all respondents interviewed by a single person systematically score too high or too low on a single measure (Bailar, 1976; Fellegi, 1974; Hansen et al., 1961; Krotki, 1978; Krotki and MacLeod, 1979).

Method effects and correlated errors

A major reason that correlated errors appear in survey research is because analysts examine multiple measures derived by the same method. When the method by which a measure was obtained affects scores on that measure (a form of measurement error that is very likely to be present to some degree), and when measures reflecting the same method effects are analyzed together, these similar method effects produce correlated errors. A brief example will illustrate the phenomenon.

Imagine a survey item that asks respondents to evaluate their own health by picking one of several answer categories ranging from "very good" to "very bad". The answers will vary, partly because people differ in the way they perceive their own health (valid variance). In addition, the answers may vary because people interpret the answer categories differently (e.g. "very good" may mean something more positive to some respondents than to others). This is measurement error attributable to method (methods variance). Now, if a second survey item using the same response scale is included in an analysis with this item on health, and if each respondent is consistent in the way he or she interprets the meaning of these categories (as could be expected), the measurement errors attributable to the method would be the same in both items. Respondents who tend to be "too high" on the first item (because of the way they interpreted the answer categories) would also tend to be "too high" on the second item, because it used the same categories. This overlap in method effects generates covariation between the items, and this covariation is added to any covariation that may exist between the concepts tapped by the items. The covariation attributable to common method effects (which is correlated error) strengthens the observed correlation if a positive relation exists among the concepts, or weakens the observed correlation if a negative relation exists among them.

Method effects can occur for any type of survey item if there can be variation in the interpretation of (or reaction to) the introduction, the question, and/or the response scale. This applies to nearly all survey items.

Method effects are not the only source of correlated errors. Correlated errors will appear whenever respondents answer two or more items, and these differences are not linked to the concept(s) the items were intended to tap. However, method effects are probably the major source of correlated error in survey data. In our own empirical investigations (described in this paper) we have not found substantial and systematic correlations among the measurement errors that could be attributed to anything other than method effects. Nor have other studies produced compelling evidence of substantial correlated errors arising from other sources, at least up to now.[6] Nevertheless, correlated error in survey data is a topic that merits further investigation.

SOURCES OF DATA

Surveys and methodological supplements

The data used in this study come from six different surveys. Five of these contacted probability samples of American adults who lived in households; the sixth was a survey of workers in a large Canadian corporation. Table 1 lists these surveys, the population each represents, the number of respondents, the method of data collection, and the substantive content of the measures used in the present investigation. The total number of respondents was 7,706 (the median number per survey was about 900; range: 376 to 3,767). Four surveys used telephone interviews to collect the data, one used face-to-face interviews, and one used a group-administered questionnaire. All were conducted as part of the regular ongoing research activity of the University of Michigan's Survey Research Center.

The basic strategy for the present investigation was to select several important concepts from the regular content of each survey, and then to add a few (6-20) additional items tapping these concepts in such a way that a multitrait multimethod data design would be achieved (Campbell and Fiske, 1959). In other words, each of the selected concepts (traits) was to be assessed by several different methods, i.e. by several distinctively different response scales. Careful attention was devoted to ensuring that the several items intended to tap a single concept all assessed exactly the same concept despite their using different response scales. The supplementary items were incorporated into the questionnaire in such a way that they constituted an integral part of the interview.[7]

[6] This is not to deny, however, that discussions of social desirability effects, yea-saying effects, and the like have suggested that these might be sources of correlated error.

[7] We expected respondents would recognize topics they had discussed previously, so explanations such as the following were sometimes included: "This next section asks about some things we have already talked about, but the questions are different. We are doing research to find the best way to ask these questions. Just tell me which anwer seems to fit your situation."

table 1: Data sources

no	date	method	N	no. of measures	concepts assessed in MTMM design
1.	Aug. 1978	pers. interv. A	3,767	19	Quality of life: assessments of housing standard of living, self, family life, community, health, life as a whole
2.	Jan. 1979	tel. interv. A	884	24	Reports about past and anticipated changes in business conditions, personal finances, health, keeping up with the news
3.	March 1979	tel. interv. A	560	9	Behavioural reports on eating too much, drinking beer, watching television
4.	July 1979	tel. interv. A	1,173	18	Reports about past and anticipated changes in business conditions, personal finances, health, keeping up with the news
5.	Sept. 1979	tel. interv. A	946	12	Quality of life: assessments of financial security, housework, health, life-as-a-whole
6.	Fall 1979	group-adm. B interview	376	24	Ratings of work organization, firm's interest in workers' welfare, improvement of working conditions, group members' knowledge about jobs, quality of groups' response, group decision making, behavioural reports of eating too much, drinking beer, watching television
	totals		7,706	106	

note: A=American adults living in households
B=Employees of Canadian business

Measures for which quality estimates were obtained

To provide a broad base for our methodological findings, the survey measures used in this study (the primary measures) are a large and intentionally heterogeneous set. In all, there are 106 primary measures that tap the 26 different concepts listed in table 1. Many of the concepts were included in two different surveys to reduce the chance that measurement conditions in any one survey would be confounded with characteristics of the concept itself.

In a further effort to broaden the base for our findings (and, at the same time, to make the measurement model estimable), a large number of different but commonly used response scales were included. In each survey, at least three different response scales were used. Across all six surveys, there was a total of 14 different scales. Table 2 provides the details. To minimize confounding the effects of response scale and survey, or response scale and concept, many of the response scales were used in more than a single survey, and all were used with several (3-8) different concepts.

table 2: Measurement methods (response scale formats)

	short name	nature of scale	survey [a]
1.	yes/no	Two categories labeled "yes", "no"	3,4
2.	better/worse-3	Three categories labeled "better", "same", "worse", plus-for some but not all measures a "don't know" category	2,4
3.	better/worse unfold	A two stage sequence in which respondent indicates in a general response (e.g. "better", "in between", or "worse") and then answer a second question to refine the position (e.g. "a lot better" or "somewhat better") or to indicate he/she hadn't "thought much about it"	2,4
4.	goodness	Five categories labeled "very good", "not very good", "not at all good"	1
5.	satisfaction	Seven categories ranging from "completely satisfied" through "neutral" to "completely dissatisfied" with unlabeled intermediate categories	1,5
6.	delighted/ terrible-7	Seven categories labeled "delighted", "pleased", "mostly satisfied", "mixed", "mostly dissatisfied", "unhappy", "terrible", plus an off-scale "no feelings at all" category	1
7.	delightful/ terrible-unfold	A two-stage sequence in which respondent first indicates a general response ("good", "bad", or "mixed") and then answers a second question to refine the position (e.g. "delighted", "pleased", "just mostly satisfied")	5
8.	ladder	Description of a ladder with 10 steps ranging from "worst feelings" to "best feelings;" respondent indicates his/her position	5
9.	graphical assessment	Line ranging from 100 (labeled "perfect") to 0 (labeled "terrible") with each decile marked	1

[a] The numbers in this column refer to the surveys listed in table 1

table 2 (continued)

short name	nature of scale	survey
10. agree/disagree	Five categories labeled "agree strongly", "agree moderately", "in the middle", "disagree moderately", "disagree strongly;" alternatively: "agree a great deal", "agree somewhat", "mixed feelings-not sure", "disagree somewhat", "disagree a great deal"	2,6
11. extent	Five categories labeled "to a very little extent", "to a little extent", "to some extent", "to a great extent", "to a very great extent"	6
12. frequency-4	Four categories labeled "almost every day", "every few days", "once or twice", "not at all"	3
13. frequency-9	Nine categories labeled "never", "hardly ever", "some of the time", "somewhat less than half", "about half the time", "somewhat more than half", "most of the time", "nearly all of the time", "all of the time", plus an off-scale "don't know" category	6
14. frequency-days	Respondent reports actual number of days per month	3

Example items and terminology

Although it is not feasible to present the exact wording and format for all 106 primary measures (these are available from the author), table 3 presents all nine primary measures from survey 3. These nine measures tap the three concepts listed for survey 3 in table 1, using the three response scales listed for survey 3 in table 2. (These nine items constituted less than 20 percent of the total interview for survey 3 and were interspersed with other material.)

In considering characteristics of items, it is useful to distinguish three distinct parts, and these can easily be seen in the example items in table 3. There may be an introduction, which is followed by one or more questions, and each question has a response scale, i.e., a set of answer categories. Together, the introduction (if any), the question, and the response scale constitute the survey item. The survey item, when answered by a set of respondents, provides a survey measure, i.e., data for analysis.

Other variables

In addition to the variables whose quality was to be assessed (the primary measures), this investigation makes use of a substantial number of other variables to account for variation in the quality assessments. These include reports by respondents about some of their own personal characteristics, ratings by interviewers about the respondents, judgments by study staff about characteristics of primary measures, and objective information about the respondents, the primary measures, and the design of the survey in which each primary measure was included.

In all, 46 such other variables were examined. These are described in this report.

table 3: The nine primary measures from survey 3

1. Now turning to things you eat and drink. Some people feel they eat too much. During the past month, how often do you feel you ate too much? Almost every day, every few days, once or twice, or not at all?
2. During the past month were there more than ten days when you drank some beer? (yes/no)
3. During the past month were there more than four days when you watched TV just to get away from the ordinary cares and problems of the day? (yes/no)

   As you know we're trying to get the most accurate information we can so I'd like to ask you about a few things we have already talked about. These may sound like the same questions, but they give you different answers to choose from. Please tell me how often each has been true for you over the past month.

4. During the past month how often did you drink beer? Almost every day, every few days, once or twice, or not at all?
5. During the past month were there more than two days when you ate to much? (yes/no)
6. How often during the past month did you watch TV just to get away from the ordinary cares and problems of the day ? Almost every day, every few days, once or twice, or not at all?

   Here are the last questions about things we asked earlier.

7. On about how many days during the past month did you drink at least one glass of beer?
8. On about how many days during the past month did you eat too much?
9. On about how many days during the past month did you watch TV just to get away from the ordinary cares and problems of the day?

ESTIMATING THE QUALITY COMPONENTS OF SURVEY MEASURES

General strategy

The measurement quality estimates were derived from a structural model of the measurement process. This model is based on a set of causal assumptions which are grounded in classic measurement theory and involve the basic notions about data quality discussed in this report.

In accord with classic measurement theory (and with what seems intuitively reasonable), a respondent's recorded answer to a survey item is assumed to reflect three types of influences:

1. the way that particular respondent feels about the concept the survey researcher intended the item to tap (e.g. the respondent's perception about changes in his or her health)
2. the way that respondent reacts to the method used for obtaining the data (particularly, in our case, the response scale)
3. everything else that might affect a recorded answer (e.g. lapses of memory by the respondent, misunderstanding by the interviewer, etc.)

A graphic portrayal of the measurement model used for survey 3 appears in figure 1. (Each of the other surveys used functionally identical models, but modified to accommodate varying numbers of concepts, response scales, and items.) As is common practice (e.g. Jöreskog and Sörbom, 1978), the exhibit shows measured variables (the primary measures) as rectangles. Unmeasured variables are shown as circles; the circles on the left represent the underlying concepts tapped by the items, and the circles on the right the response-scale-based sources of measurement error. One-way arrows indicate direct causal influences; two-headed arrows indicate relationships between concepts where no assumption about causality is made. (The shortest one-way arrows in figure 1 allow for residual variance to influence the observed measures.) Technically, this is a confirmatory factor analysis model and could be expressed as nine simultaneous equations.[8]

If one has confidence in such a model (a topic addressed below) some of the parameters can be interpreted as measurement quality assessments because they indicate the extent to which variation in a given measure reflects:

[8] Texts and manuals on structural modeling (e.g. Jöreskog and Sörbom, 1978) describe procedures for writing such sets of equations.

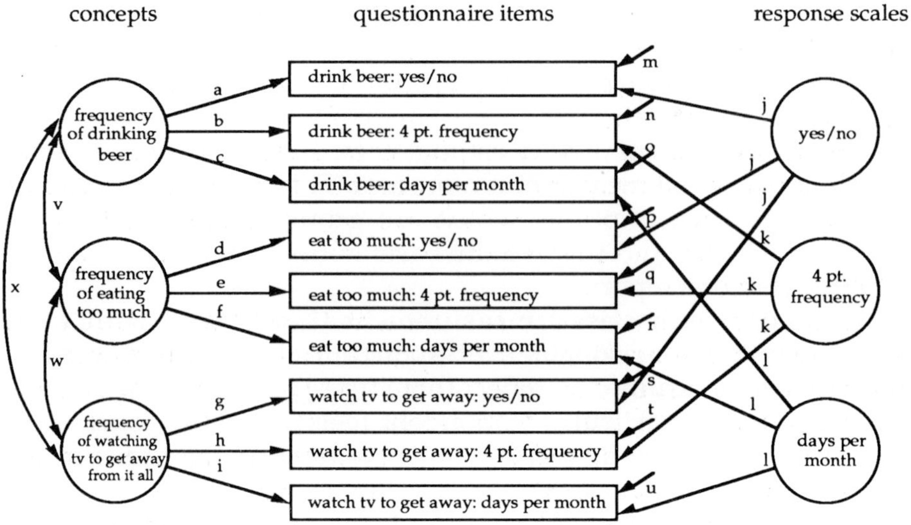

figure 1: Schematic form of structural model

1. differences among respondents with regard to the underlying concept, an estimate of construct validity (parameters a-i in figure 1)
2. differences among respondents in the way they interpret and use the response scale, a major source of correlated error (parameters j-l)
3. the extent to which the measure reflects other influences, primarily random measurement error (parameters m-u). Other parameters can be interpreted as estimates of the true relationships among the concepts (parameters v-x).

The process of estimating the structural model involves finding a unique set of strengths for the causal linkages (parameters a-l, imputed correlations among latent concepts (parameters v-x), and variances for the residual error sources (parameters m-u) that will produce a set of predicted relationships among the observed measures that come as close as possible to the observed relationships among these measures. In this study these parameter values were obtained by the maximum-likelihood estimation technique incorporated in the LISREL computer program (Jöreskog, 1978; Jöreskog and Sörbom, 1978).[9]

[9] As indicated in figure 1, an equality constraint was imposed on each set of method effect parameters (the sets of j, k and l parameters), and each method effect was constrained to be independent of all other latent variables. These constraints ensured that the model's estimates

Generating the measurement quality estimates that are analyzed in this report required running 125 measurement models. Each of the six surveys involved a series of models of a given form, and within each series, runs were made for different (and sometimes overlapping) groups of respondents. The nature of the model applied to each survey can be determined from the information available in tables 1 and 2, as has been illustrated for survey 3.[10]

Relevant precedents

Although this study represents the first broad scale attempt to use structural modeling technology to generate measurement quality estimated for data from regular on-going surveys, this investigation is grounded on prior work.

The basic idea of using a multitrait multimethod data design for assessing validity and method effects was proposed by Campbell and Fiske (1959). As the potential of structural modeling for handling multitrait multimethod data become recognized, many investigators advocated its use (e.g. Alwin, 1974; Boruch et al., 1970; Schmitt et al., 1977). The general usefulness of structural models for illuminating the nature and quality of social data now seems well established, and measurement models receive extensive attention in a number of recent texts on social measurement and analysis (e.g. Zeller and Carmines, 1980; Bohrnstedt and Borgatta, 1981; Sullivan and Feldman, 1979; Namboodiri et al., 1975).

Although theoretical and methodological discussions about the use of measurement models are no longer rare, only recently have investigations that actually use the new modeling technology to estimate data quality begun to appear (Andrews, 1979; Andrews and Crandall, 1976; Mason et al., 1976; Kluegel et al., 1977; Robins and West, 1977; Bielby and Hauser, 1977; Bielby et al., 1977a, 1977b; Mare and Mason, 1980; and Corcoran, 1980.)

of method effects reflected the assumptions that each method effect was a statistically unique phenomenon and that it had an equal impact on all measures based on that method. These constraints also helped to identify the other parameters in the model. (The fact that the models estimated for this study fit as well as they did suggests that it was not unreasonable to assume that the response scale factors were statistically independent of one another and of the substantive concepts being measured. This result is consistent with those from a previous study (Andrews and Crandall, 1976), where relationships among response scale factors were left unconstrained and empirical estimates showed them to be very weak).

[10] This same approach can be applied straighforwardly to surveys 2, 5, and 6, which involve complete 3 x 8, 3 x 4, and 3 x 8 method-by-trait designs, respectively. Surveys 1 and 4 are complicated by the fact that the multitrait multimethod data design was not complete, i.e. some concepts were not assessed by all methods.

REASONS FOR CONFIDENCE IN THE ESTIMATES

Beyond the fact that other investigators are using and advocating the same general approach as has been used in this study, what empirical evidence is there that suggests that the results from our modeling analyses provide reasonable estimates of data quality?

Appropriateness of the data for the measurement model

The measurement model assumes the data can be adequately explained within the confines of an additive linear system. Furthermore, given that product-moment correlations were input to the LISREL computer program, they are assumed to appropriately represent the actual relationships in the data.[11] To see wether these assumptions were met, the data from each survey were scanned for instances of marked skews, nonlinearities, and non-additives. In most cases we concluded that the data in their original form were appropriate for analysis; in a few instances a transformation was effected to reduce skew and/or curvilinearity before the correlations were computed.

Adequacy of the measurement model

Models that work well are theoretically reasonable, manage to closely account for the observed relationships in the data, and do so with parameters that are themselves of reasonable magnitude. How well did the models used in this study rate on these criteria?

The theoretical relevance of the model has already been discussed. The model represents a direct implementation of classic measurement theory, and the primary measures (those represented in the multitrait multimethod matrices) were designed specifically for use with this model.

Models of the form displayed in figure 1 proved to fit the data well, i.e., they did a good job in accounting for the actual relationships among the measures. In over a hundred runs involving the application of this model to different sets of variables and/or different sets of respondents, this model consistently produced predicted relationships that tended to be close to the observed relationships. The mean absolute deviation between predicted and

[11] Readers who are sophisticated with regard to structural modeling may wonder how the use of correlations as input to the model interacted with the equality constraint imposed on each set of method effect parameters (the only reason that it would make any difference wether input was correlations or covariances). A check showed that all measures obtained using any one method tended to have similar variances, and hence it made no difference wether correlations or covariances were used as input to the models.

observed correlations was always less than .10, and in most runs it was less than .05.[12]

Although a very wide range of parameter values was obtained, in nearly all cases these values were reasonable in the light of theoretical expectation. Specifically, validities, method effects, and residual errors should all be within 0 to 1. Although there was no constraint on the estimation of parameters to prevent the occurrence of unreasonable values, they very rarely occurred. Out of 2,115 validity estimates that were generated, none was less than 0 and only 19 (0.9 percent) exceeded 1.00 (the highest was 1.13). Of 2,115 method effect coefficients, none was less than 0 and none exceeded 1.00. Of 2,115 residual error coefficients, 32 (1.5 percent) were less than 0 and none was greater than 1. In short, out of more than 6,000 measurement quality estimates, 99.2 percent had reasonable values, and the few that fell outside the reasonable range were not far outside.[13] Thus, on the criteria of theoretical reasonableness, adequacy of fit, and reasonableness of estimates, the model seems good. Of course, this is not to say that this model always produced the best possible fit. In many cases, minor modifications could be made that would modestly improve the fit. However, despite considerable searching, we have been unable to find any way to change the model that would produce consistent and substantial improvements in fit.[14]

[12] This examination of the congruence between observed and predictive relationships, as well as our general approach to the topic of model fit, is accord with perspectives advocated by Bentler and Bonett (1980) in an article on assessing how well models fit data. Their article was not published until after all computing for this study had been completed, but had we been able to compute the fit indices they propose for each of the 125 models we ran, there is little doubt that the indices would show that our models account for most of what is going on in the data.

One general indicator of model fit is the ratio of chi square to the degree of freedom. The median value of this ratio over the 125 models was 1.87 (mean value = 2.24). These low values are another indication of the satisfactory fits obtained.

[13] One might ask why any unreasonable estimates were generated. Our guess is that they result from minor inconsistencies in the input correlations arising from some respondents being omitted from the calculation of one correlation (because their data were missing on one or both of those particular variables) while different respondents were omitted from other correlations. Although the use of pairwise missing data deletion from correlation matrices has this drawback, it generally leads to less serious problems than what can occur with casewise deletion, in which a great many respondents can be lost from the entire matrix.

[14] One exploration took advantage of the fact that all of the respondents to the July survey had also participated six months earlier in the January survey, i.e. this pair of surveys constitutes a two-wave panel. The data were collected in this way specifically to allow investigation of wether the residual variance really looks like random error, or wether there is some part of it which, while independent from that of any other measure assessed in any one survey, might be a reliable aspect of that particular item. Results of this panel analysis were clear: there was little, if any, reliable unique variance for any of the 12 measures that had been repeated. (The estimates of the reliable unique variance components ranged from 0 percent to 11 percent with

CHAPTER 1

ANALYSES OF THE MEASUREMENT QUALITY ESTIMATES

Analyses strategy

The preceding stages of this research resulted in the generation of a large number of measurement quality estimates. Specifically, there were 2,115 sets of estimates, where each set consists of a validity estimate, a method effect estimate, and a residual error estimate for a particular survey item as it was answered by a particular group of respondents on a particular occasion. The next step in the research was to analyze these quality estimates to determine under what conditions they tended to be higher and under what conditions lower.

This was done in three steps. First, simple descriptive statistics showing the level and variation of these measurement quality estimates were examined. Second, the effects of survey design characteristics were examined. A series of multivariate analyses explored how aspects of survey design (characteristics of response scales, item wordings, the topic investigated, and questionnaire design) related to each of the three measurement quality estimates. Third, after the effects of survey design characteristics (age, education, sex, and many others) were examined. This strategy provides a direct way of addressing some of the key questions in survey research: how to achieve more valid measures. The approach, however, is unusual. Few previous studies have had a sufficient number of measurement quality estimates, together with information about the survey and respondent characteristics associated with each estimate, to make a direct multivariate analysis of those estimates feasible.

It is important to note that analysis strategy involves a shift in data bases. When the measurement quality estimates were being generated, the "cases" were survey-items-as-administered-to-a-particular-set-of-respondents.[15] For example, one of the "cases" in this data set is a survey item about satisfaction with own health that was answered on a 7-point scale of satisfaction in August 1978 by a representative sample of Americans 71 to 90 years old. This is just one among more than 2,000 such cases.

a mean of 2 percent.) This is further important evidence that it seems appropriate to use the residual parameters in our measurement models as estimates of random error.

15 Creation of this new data set required transcription of a large number of LISREL-produced measurement quality estimates and assembly of many descriptive characteristics for each survey item. These data were then puched, verified, and built into a computer data file. The reliability of this coding process was checked for a sample of the cases. The coding accuracy was found to be 99.4%, a level that was judged to be highly satisfactory.

Data quality level and variation

Table 4 presents basic descriptive information about the measurement quality levels, and the variation in those levels, observed in this research. Presented there are the univariate statistics for all 2,115 estimates of each of the three measurement quality components that will be "dependent variables" in this analysis.[16] These results are of considerable interest in their own right.

The first row in table 4 presents the mean for the estimates of validity, method effect, and residual error. Note that the average validity is .81, the average method effect is .16, and the average residual error effect is .53. These figures, when squared, indicate that the "typical" survey measure examined in this research consisted of 66 percent valid variance. (These sum to 97 percent. The discrepancy from 100 percent presumably reflects minor imperfections in the way the measurement model used to generate the quality estimates fit some of the data). Although no claim is made that the set of survey items examined here is representative of all items used in current surveys, this set of items is broader and more heterogeneous than any other we know of whose quality has been estimated, and hence these estimates for the quality of our "typical" item probably provide the best available information about these aspects of measurement quality for single-item survey measures tapping rather specific attitudes and behaviours.[17]

Table 4 also presents the standard deviations of the quality estimates. These range from .10 to .16. For the validity estimates, which have a standard deviation of .10, one can infer that about two-thirds of all validity estimates fell in the range .71 to .91.Hence, about two-thirds of the survey measures examined here contained between 50 percent and 83 percent valid variance, and about one-sixth had less than 50 percent valid variance.

Comparable calculations for method effects suggest that two-thirds of the measures had between 0 percent and 7 percent method variance. Similarly, two-thirds of the measures had between 14 percent and 48 percent residual variance.

[16] The measurement quality estimates are taken directly from the structural modeling analyses described in this paper. The validity and method effect estimates are the LISREL-produced λ parameters, and the residual error estimate is the square root of the LISREL-produced θ parameter. The square root transformation of the θ parameter was used so all three measurement quality estimates would be on the same scale, i.e. the square of each of them indicates the percentage of variance of the indicated kind in the survey measure.

[17] The median quality estimates in the second row of table 4, and the following skew and kurtosis coefficients for these quality estimates, confirm that their distributions are approximately normal. The skew coefficients for the estimates of validity, method effects, and residual error are, respectively: -.48, .21, and -.77; kurtosis coefficients are: .42, -.46, and 1.14.

table 4: Measurement quality estimates, average level and variation

	validity	method effect	residual error
mean	.81	.16	.53
median	.81	.16	.55
standard deviation	.10	.11	.16
number of estimates	2,115	2,115	2,115

note: The estimates of validity and method effects are based on λ parameters from LISREL; the estimates of residual error are based on the square roots of the θ parameters from LISREL.

SURVEY DESIGN CHARACTERISTICS AND DATA QUALITY

The next step of the analysis was to perform a series of multivariate analyses to attempt to explain the observed variation in the quality estimates. Given the nature of the available predictor variables and the lack of previous exploration in this area, it was important that the multivariate analyses be able to incorporate nominal-scale predictors,nonlinear relationships, and interactive (i.e., nonadditive) effects. Accordingly, multiple classification analysis using pattern variables (Andrews et al., 1973) and SEARCH (formerly known as AID (Sonquist et al., 1974; Survey Research Center Computer Support Group, 1981)) were selected as the primary multivariate analysis methods.

Predictive power of survey characteristics

After considerable exploration using various combinations of the available independent variables, a final multivariate analysis was selected that used 13 predictors.[18] (One of these predictors was a pattern variable based on two more variables). These 13 predictors are listed in table 5, together with an indication of the explanatory

[18] Key questions addressed in these preliminary analyses had to do with (1) an appropriate and feasible set of predictors for the MCA, (2) wether important interactions were present that, without special treatment, would be unrepresented in the additieve MCA model, and (3) wether the predictors were defined in a sufficiently general way that they could be used in future studies of measurement quality. In addition to the survey characteristic variables that finally entered the analysis, 11 others were also examined. (There were, as expected, certain combinations of survey design characteristics that were perfectly or nearly perfectly confounded with others. These had to be indentified and handled in some way- by eliminating one of the potential predictors from the analysis or by conbining appropriate categories). In addition, a series of SEARCH runs was helpful in selecting the more useful predictors when some had to be omitted and in assuring that no major interactions were being missed. Given that the proportions of variance explained in the final SEARCH runs were extremely close to the R^2's achieved by the multiple classification analyses (presented in table 5), we can be confident that the MCA model is appropriate for these data.

power of each, both single and in combination with all others, and with an indication of the total explanatory power achieved by the entire set.

Table 6 presents more detailed results from this analysis and shows the effect on each of the three measurement quality estimates of each category of each predictor variable, both before and after holding constant the other predictors.

One of the first things to note in table 5 is that the characteristics of survey design represented in this analysis account for a large part of the variation in the estimates of validity, method effects, and residual error. As shown by the adjusted R^2's, the survey design characteristics account for 66 percent to 72 percent of the variance in the dependent variables. This is an important finding, for it shows that much of the variation in measurement quality can be explained, and hence measurement quality is subject to prediction in other surveys and perhaps to improvement.

As indicated in table 5, the characteristics of survey design have been grouped into four conceptually distinct sets. The first includes four aspects of the response scale used with the survey item. Next comes a set of variables that tap several characteristics of the survey item itself. The third set has to do with the design of the questionnaire or interview schedule and the data collection mode. The final set of survey characteristics includes four variables tapping various aspects of the topical content of the survey item.[19]

The β^2's in table 5 indicate the relative importance of the various survey characteristics in accounting for validity, method effects, and residual error. (The η^2's, which reflect the simple bivariate relationships, are also interesting, but the multivariate results (reflected in the β^2's) are the more useful.) The most important aspects of survey design, as indicated by average β^2's of .25 or more, are the number of answer categories in the response scale, wether these answer alternatives include a "don't know" category, battery length, and wether the item uses an absolute or comparative perspective.

Making appropriate choices with respect to these design characteristics can apparently, have an important impact on the measurement quality in a survey. Survey characteristics with a more moderate and/or less general effect (average β^2's in the range .15 to .24) include wether answer categories are all labeled, the length of the question and its introduction, and the position of an item in the questionnaire.

[19] These four variables were based on ratings made by project staff members.

table 5: Summary results from multiple classification analysis using 13 aspects of survey design to predict validity, method effects, and residual error

	quality component being predicted						
	validity		meth. eff.		res. error		
	adj. η^2	MCA β^2	adj. η^2	MCA β^2	adj. η^2	MCA β^2	mean β^2
characteristics of response scale:							
number of scale categories	.19	.56	.20	.68	.25	.74	.66
explicit: "don't know" option	.06	.31	.09	.45	.02	.30	.35
category labeling	.00	.27	.04	.28	.00	.15	.23
explicit midpoint	.01	.01	.03	.06	.01	.00	.02
characteristics of the item:							
absolute vs. comparative	.00	.28	.15	.15	.02	.33	.25
length of intro. & question	.12	.13	.27	.35	.12	.10	.19
questionnaire design, data collection:							
battery length	.10	.17	.09	.19	.13	.44	.27
position of item in questionnaire	.24	.13	.22	.18	.24	.16	.16
data collection procedure	.02	.03	.18	.24	.01	.02	.10
characteristics of the topic:							
sensitivity to social desirability	.09	.07	.04	.00	.11	.08	.05
content specificity	.06	.06	.04	.00	.07	.04	.03
experience versus prediction	.04	.01	.04	.00	.05	.01	.01
content salience	.08	.01	.09	.00	.08	.00	.00
joint explanatory power of 13 predictors (R^2 adj.)	.66		.72		.67		
N (number of quality estimates)	2,115		2,115		2,115		

notes: η^2, the squared correlation ratio, shows the proportion of variance in a dependent variable explained by one predictor variable considered alone. The coefficients shown here include an adjustment for shrinkage likely to occur upon replication.
MCA β^2 is a measure of the strength of relationship between a dependent variable and a predictor while holding constant the effects of all other predictors included in this analysis.
R^2, the squared multiple correlation coefficient, shows the proportion of variance in a dependent variable explained by all predictors jointly. An adjustment for likely shrinkage upon replication has been incorporated.

Choices with respect to these matters can also have important effects on measurement quality. Equally interesting are the design characteristics that did not have substantial effects on data quality. These include all four aspects of the topic being asked about, wether the answer scale includes an explicit midpoint, and the data collection procedure. The effects of some of these design matters have been the subject of considerable debate among survey researchers, and it is of interest to find that (at least in this study) they have relatively little impact on measurement quality.

Specific effects of survey design

In table 6 one can see the way each of these characteristics of survey design relates to validity, method effects, and residual error. Presented there are the effects of each category of each predictor variable on each component of measurement quality, both before and after controlling for the effects of the other predictors. The coefficients in the biv. columns are results from the simple bivariate analysis (the relationships that are summarized by the η^2's in table 5), and the coefficients in mult. columns come from the multiple classification analyses (and are summarized by the β^2's in table 5). As noted earlier, the multivariate analysis results probably are the more useful.

The multivariate coefficients in table 6 show the amount by which the quality component would go up or down from the mean (presented in table 4) if a measure had the characteristic indicated and there were no effects from any of the other predictor variables. For example, the -.06 effect on validity of using an item with a 2-point answer scale means that, holding everything else constant, validities of such items can be expected to be 6 "points" lower than that of the average item, i.e. .75 (=.81 - .06). Although the coefficients presented in table 6 may appear small to the uninitiated reader, in many cases they show sharp and important effects.

Number of scale categories

The number of answer scale categories is shown in table 5 to have the biggest effects on data quality, and the multivariate coefficients in table 6 show that, in general, as the number of answer categories goes up, data quality goes up, i.e. validity tends to increase and residual error tends to decrease. (The trend for method effects is less clear, though as its extremes it follows the general trend.) The validity and residual errors results show an interesting and possibly important curvilearity at the low end of the scale: both 2-point and 3-point scales give less good measurement quality than scales with four or more categories, but 2-point scales are not as bad as 3-point scales.

These results are consistent with a wide-ranging and uncoordinated literature showing that use of more categories (at least up to 507) produces a more accurate reflection of the underlying vari-table 6: Effects of survey design characteristics on data quality[a]

	no. of estimates	validity biv.	validity mult.	meth. eff. biv.	meth. eff. mult.	res. error biv.	res. error mult.
CHARACTERISTICS OF THE RESPONSE SCALE							
number of scale categories:							
2	120	-.00	-.06	.04	.11	.02	.04
3	364	-.09	-.13	-.07	-.05	.14	.22
4-5	542	.01	.04	.00	.01	-.00	-.06
7	650	.02	.00	-.03	-.02	-.01	.04
9-19	208	.04	.01	.09	.21	-.09	-.07
20+ or actual frequency	231	.05	.14	.07	-.13	-.14	-.28
explicit "don't know" option:							
no	1,516	-.02	-.04	.02	.04	.01	.06
yes	599	.04	.09	-.05	-.11	-.03	-.14
category labeling:							
all categories labeled	1,502	-.00	-.04	-.01	.04	.00	.03
some categories unlabeled	613	.00	.08	.03	-.09	-.01	-.10
explicit midpoint:							
no	532	.02	.02	-.03	.05	-.03	.01
yes	1,583	-.01	-.01	.01	-.02	.01	-.01
CHARACTERISTICS OF THE ITEM							
absolute versus comparative:							
absolute	1,275	.00	-.04	.03	.01	-.02	.07
comparative	840	-.01	.07	-.05	-.02	.03	-.11
length of intro. and question:							
short intro., short question	56	-.02	-.05	-.04	-.01	-.09	-.02
short intro., med. question	231	-.03	-.05	-.03	.03	.06	.08
short intro., long question	259	-.06	-.01	-.04	-.03	.09	.05
medium intro, short question	249	-.03	.03	-.01	-.08	.05	-.04
medium intro, med. question	365	-.02	.06	-.00	-.08	-.01	-.07
medium intro., long question	44	.12	.06	-.02	-.09	-.16	-.05
long intro., short question	351	.00	-.00	.11	.05	-.04	-.00
long intro., medium question	219	.03	.00	.02	.07	-.05	-.00
long intro., long question	341	.03	-.03	-.06	.07	-.19	.06

note: The estimates of validity and method effects are based on λ parameters from LISREL; the estimates of residual error are based on the square roots of the θ parameters from LISREL.
[a] Bivariate coefficients show deviations from the mean that are associated with membership in the designated category; multivariate coefficients are similar except effects of other predictors have been held constant by multiple classification analysis.

table 6 (continued)

	no. of	validity		meth. eff.		res. error	
	estimates	biv.	mult.	biv.	mult.	biv.	mult.
QUESTIONNAIRE DESIGN							
data collection procedure:							
telephone interview	1,332	.01	-.01	-.03	-.02	-.02	.01
face-to-face interview	399	-.02	.00	.09	.10	.03	-.01
group administered questionnaire	384	.01	.04	-.01	-.05	.02	-.04
battery length:							
1 item (i.e., not in battery)	537	-.05	.03	-.03	-.04	.08	-.10
2-4 items	519	.03	.05	.03	.02	-.08	-.11
5-9 items	736	.02	-.03	-.03	-.03	-.02	.09
10 or more items	323	-.00	-.07	.04	.10	.02	.14
position of item in questionnaire:							
1-5 i.e., among first five items)	148	-.09	-.01	-.06	.03	.15	.03
6-25	280	-.09	-.02	-.03	-.02	.13	.05
26-35	336	.01	.04	.00	-.01	-.01	-.07
36-39	276	.04	.02	-.07	-.03	-.04	-.03
40-100	493	.05	.04	.00	-.05	-.10	-.06
101-200	309	.02	-.04	-.00	.07	-.01	.08
201-348	273	-.02	-.06	.11	.06	.01	.08
CHARACTERISTICS OF THE TOPIC							
sensitivity to social desirability:							
low or medium	1,587	-.02	-.02	.01	.00	.03	.03
high	528	.05	.05	-.04	-.00	-.09	-.08
content specificity:							
low	700	-.02	-.03	-.01	.01	.04	.04
medium	401	-.02	-.01	.04	-.00	.04	.02
high	1,014	.03	.02	-.01	-.01	-.04	-.03
experience versus prediction:							
actual experience	1,753	.01	.00	.01	-.00	-.02	-.01
predictions	362	-.04	-.02	-.05	-.00	.08	.04
content salience:							
low	445	-.05	.01	-.02	-.01	.08	-.01
medium	1,056	.03	.01	-.02	-.00	-.03	.00
high	614	-.01	-.01	.05	.01	.00	.01

statistical significance:
By conventional tests of significance, a difference between most pairs of multivariate coefficients of .02 or more is significant at the p=.05 level. Standard errors for the multivariate coefficients (Hill, 1979) range from about .002 when N is 800 to about .01 when N is 50.

ation. See, for example, Bollen and Barb (1981), Cochran (1968), Conner (1972), Cox (1980), Lissitz and Greene (1975), Martin (1978), Pearson (1913), and Ramsey (1973). This previous body of literature, however, would not have predicted the marked superiority of our "20+" category, and it is possible that this aspect of our results is an artifact (for the reasons in footnote 21).

Explicit "don't know" option

According to table 5, the second most important survey characteristic is wether the answer categories include an explicit "don't know" option. The effect of this design matter is clear and consistent: Inclusion of an explicit "don't know" category was associated with better data, higher validity, lower method effects, and lower residual error. The reasonable idea that one should let respondents "opt out" if they lack the requisite information or opinions receives strong endorsement.

Battery length

The third most important predictor denotes wether the item is included in a battery with other items, and if so, how long that battery is. The primary effect here is with respect to residual error, but the results for validity and for method effects, though weaker, tend to follow the same trend. The results show that the longer the battery, the lower the data quality.

Including items in batteries where all share a common introduction and/or identical answer scale offers obvious advantages with regard to efficiency and speed, but these results suggest that such gains come at the cost of reduced measurement quality if the battery consists of more than just a few items. It is possible that respondents and/ or interviewers recognize the "production line" character of this survey strategy and that it promotes carelessness in the way questions are asked and answered.

One might have guessed that what mattered would not be the total length of the battery in which an item is included but rather how far into a battery the item is located. Both characteristics of items were examined, and, to our surprise, battery length showed stronger relationships to each of the measurement quality components than did position in battery. By logical necessity, the two variables were highly correlated, and it was not feasible to retain both in the multivariate analysis.

Absolute versus comparative

The fourth most important predictor taps wether the item uses an absolute or comparative perspective. The results in table 6 clearly favour the comparative approach. It may be that the provision of some "anchor points", as is required in the comparative approach, helps respondents give more precise answers.

Length of introduction and question

Two item-design characteristics, the length of the introduction to a question and the length of the question itself, were combined into the single nine-category pattern variable that was used in the multiple classification analysis because preliminary exploration had shown these characteristics to have both curvilinear main effects and a first-order interaction.[20] Table 6 shows that validity tended to be highest and both types of error lowest when question were preceded by a medium length introduction (defined as an introduction of 16 to 64 words). Furthermore, given a medium length introduction, medium or long questions (16-24, or 25+ words, respectively) yielded higher data quality than shorter questions. The overall pattern of these results suggests that short introductions followed by short questions are not good (perhaps the respondents does not have an opportunity to get a clear understanding of what is being asked and/or does not have time to develop a precise answer) and neither are long introductions followed by long questions (respondents may lose track of what is being asked and/or get bored while waiting to answer).

Position of item in questionnaire

Table 6 shows a consistent moderate-strength tendency for data quality to be lower when items were at the beginning of a questionnaire (within the first 25 items) or far into a long questionnaire (beyond the 100th item). Better data quality comes from items that fell in the 26th to 100th positions.

It is not hard to imagine how this effect might come about. Items that come early in a questionnaire may be presented before the respondents is "warmed up" to the task and, in an interview, before rapport between interviewer and respondent has been developed. On the other hand, after the 100th item, respondents and/or interviewers may begin to suffer from fatigue or become careless.

[20] Andrews et al. (1973) describe how interactions can be handled in a multiple classification analysis through the use of pattern variables.

Category labeling

The moderate-strength relationships associated with category labeling were a surprise and are not yet fully understood. The contrast is between items whose answer categories were fully labeled, i.e. an explicit meaning was indicated for every possible answer, and items where some of the answer categories were left unlabeled, as in a format where only the end points are labeled and some intermediate points take their meaning from their relative position on the page. Contrary to expectation, the results of the multivariate analysis suggest that data quality is below average when all categories are labeled.[21]

Predictors showing weak links to quality

All of the remaining predictors have average β's of .10 or less, and with just a single exception none of the individual β's associated with these predictors exceeds that level.[22] The fact that some of these relationships between survey design and data quality are weak is of great interest. It is helpful to know that the often debated issue of wether to allow respondents an "easy out" by including an explicit midpoint ("neutral", "pro-con", etc.) had only slight effects on data quality. It is also interesting to observe that it made little difference wether an item asked about things the respondent had already experienced or asked for predictions about the future. (The small effects that do appear here are in the expected direction, i.e. favouring things the respondent has experienced.) And the finding of only very small effects on validity and residual error attributable to wether data were collected by telephone interviews, face-to-face interviews, or group-administered questionnaires will also be encouraging to many survey researchers. That none of the characteristics of the substantive topic being asked about in a survey item had an important effect on data quality was in some respects a disappointment, for we had clear expectations about how these variables might relate. However, the fact that the relationships all proved very weak might be seen as a desirable outcome by many, for it suggests that (other things

[21] It is possible that the surprisingly good quality associated with answer scales having 20+ categories (many of which consisted of reports of actual frequencies per month of doing various things, and hence which count as "fully labeled") has interacted with the category labeling predictor so as to produce an over estimate of the quality of data from 20+ answer categories and a corresponding under estimate of the data from fully labeled categories. Unfortunately, the data are not sufficient to clarify this matter and it merits further investigation.

[22] The exception is an indication that face-to-face interviews result in a slight enhancement of method effects, but this is not accompanied by any reduction in validity for measures obtained in face-to-face interviews.

equal) data of at least average quality can be obtained about a wide range of topics.

RESPONDENT CHARACTERISTICS AND DATA QUALITY

The idea that characteristics of respondents may relate to data quality is an appealing one and can be readily investigated in a sophisticated way with these data. This analysis is possible because estimates of data quality were obtained for many different subgroups of respondents. For example, data quality estimates were obtained separately for young respondents, middle-aged respondents, and elderly respondents, and hence it is possible to see how validity, method effects, and residual error (each averaged across many survey items) varies with age.[23]

Analysis strategy

Because the respondent subgroups for which data quality estimates could be obtained varied from survey to survey, and because survey design characteristics (e.g. topics investigated, answer formats used, data collection procedures, etc.) also varied from survey to survey, and because survey design characteristics have a major impact on data quality (as discussed in the preceding section) it is necessary to remove the survey design effects before looking at the effects of respondent characteristics. This was accomplished through residualization.

The dependent variables in the analyses to be described are the deviations of the actual estimates of an item's validity, method effects, or residual error. from what would be predicted to be that item's measurement quality given the design of the survey in which that item occurred. Each of these residualized scores were then related to a 53-category predictor variable that denoted the group of respondents on whom the estimates of measurement quality were calculated.[24] These respondent characteristics explained about 12 percent of the remaining variance in the validity estimates, 16 percent for the method effects, and 5 percent for residual error. Clearly, respondent characteristics were not a major predictor of variation in the quality of measurement in these data, but there are some characteristics that showed interesting and potentially important links to measurement quality.

[23] A separate paper (Andrews and Herzog, 1982) provides a detailed analysis of the relationship between survey data quality and respondent age.

[24] Space limitations permit only a summary of this analysis here. Full details appear in Andrews (1984).

Specific effects of respondent characteristics

Validity was higher for more educated respondents, for younger respondents, and for whites. Residual error showed exactly the opposite trends. (Method effects had a clear trend only with respect to age, where it showed sharp increases with increasing age.) Of course these demographic characteristics are known to be substantially correlated among United States adults, and an immediate question is wether these trends would hold up under various controls. A series of subsidiary analyses (not shown) indicated that most of the race effects disappeared when education was controlled. Apparently they were attributable to the fact that blacks tend to have less education than whites. However, the age and education effects persisted even when the other variable was controlled.

There were virtually no relationships between the sex of the respondent and the indicators of data quality. Interviewers' impressions about respondents' interest in the survey topics, general intelligence, sincerity with which questions were answered, suspiciousness, and reluctance to participate showed effects in the expected direction (i.e. high interest, intelligence, and/or sincerity tended to be associated with higher validity and lower errors) but none of the effects was very large. There was also evidence that respondents who had difficulty with the interview (e.g. received assistance provided by the interviewer, made requests for clarification and/ or repetition of questions) tended to give answers with lower validity and higher residual error.

A persistent concern of survey organizations is wether it is "worth it" to try to contact hard-to-reach respondents. In this analysis no quality differences were observed between respondents contacted on the first attempt and those contacted only after five or more attempts. There also were no quality differences related to wether the respondent had participated in a previous survey conducted by the Survey Research Center.

In a few surveys we included five items selected from the Crowne-Marlowe (1964) scale of social desirability in order to obtain an indication of respondents' concern for presenting themselves in socially desirable ways. This scale produced results in the expected direction: respondents who scored relatively high on this concern had a modest tendency to give data that were below average in validity and above average in residual error.

Further explorations

A legitimate concern is wether the effects on data quality of the various survey design characteristics explored in the previous section are the same for all types of respondents. In formal statistical terms, are there interactions involving respondent characteristics, survey design characteristics, and data quality? With 22 sets of contrasting respondent characteristics, 13 sets of survey design characteristics, and 3 data quality assessments, there were over 800 first-order interactions that might potentially occur. Eight of these that promised to be most interesting, and for which data were available in sufficient depth, were checked in detail. These included combinations of age, education, or respondent concern for social desirability with the number of answer categories, data collection procedure, length of introduction and question, or item sensitivity to social desirability, as related to the mean levels of validity. The general results of this exploration was that no major interactions were found. While we cannot, of course, be sure that interactions do not exist where we have not checked for them, we have increased confidence that most of the survey design effects described in the preceding section will be generally applicable to a wide range of different types of respondents.

IMPLICATIONS OF THE STUDY

There are at least four ways in which the outcomes of this investigation may prove useful.

IMPLEMENTING THE TECHNIQUE

One of the most important outcomes of this research is the discovery that it was indeed feasible to generate measurement quality estimates using the new measurement modelling techniques in regular on going surveys. The marginal cost of obtaining measurement quality estimates for key concepts assessed in a survey can be quite low. Since the number of additional items is not large (rarely more than 10-20), this is not a major cost. The other component of cost arises from the staff time required for the initial design of the data collection and from the staff time and computing charges for the subsequent analysis of those data. For a professional with the requisite skills (or with access to skilled consultants), neither of these need be a major task.

RECOMMENDATIONS ABOUT SURVEY DESIGN

A second general implication of this investigation is the promise it holds for generating knowledge about how to design surveys that will yield higher quality measures. The multivariate analysis described in this report itself includes many suggestions for ways to enhance measurement quality, and it could be a prototype for similar analyses performed on other measurement quality estimates that may become available for other survey measures, for other national or cultural settings, and/or for specialized groups of respondents.

PREDICTING THE MEASUREMENT QUALITY OF OTHER ITEMS

One of the potential uses for the detailed set of coefficients presented in table 6 is to generate predictions of measurement quality for survey items that were not actually included in this study. If it were not feasible for a survey team to develop its own multitrait multimethod methodological supplement and generate its own estimates of measurement quality, estimates could be obtained by extrapolating the results from the present investigation. How accurate such estimates would be is, of course, open to question. However, almost certainly it would be better to use them than to totally disregard measurement error. The predictions derived by extrapolating from results in this study should be more accurate as:

1. the items whose measurement qualities are being predicted are more similar in type of content and format to the items examined here
2. the surveyed population is more similar to a general American or Canadian adult population
3. the surveying organization uses methods and procedures more similar to those of the University of Michigan's Survey Research Center.

To make a prediction of the measurement quality of a survey item, one would begin with the means shown in table 4. These would be adjusted upward or downward according to the particular combination of survey design characteristics that pertain to the item for which one is making the prediction. The appropriate adjustment is determined by adding together relevant coefficients selected from table 6. Of the 13 survey design characteristics presented in table 6, all except the final four should be easy to determine for any item. (One would not go very astray if the final four were simply neglected, because none of them correlates very strongly with other predictors and none has a strong effect on any of the measurement quality estimates.)

MEASUREMENT QUALITY AND OBSERVED RELATIONSHIPS

This report began by observing that measurement errors influence observed relationships. It is appropriate to conclude with a brief discussion of how one can use information about measurement error to make inferences about the true relationships.

Bivariate relationship between single-item measures

The basic assumptions of measurement modeling, which can be represented in the algebra of path analysis (Zeller and Carmines, 1980), predict that a simple observed relationship will be equal to the true relationship between the concepts being tapped times the product of the validities of the measures, plus the proportion of correlated error in the measures. Algebraically,

$$r_{AB} = r'_{AB} V_A V_B + E^c_{AB} \qquad (1)$$

where

r_{AB} is the observed product-moment correlation between measures A and B

r'_{AB} is the true correlation between the concepts tapped by measures A and B

V_A is the construct validity of measure A, i.e., the square root of the percentage of valid variance

V_B is the construct validity of measure B

E^c_{AB} is the correlated error shared by measures A and B, i.e. the product of the common method effect coefficients for variables A and B.

This formula can be transformed to provide predictions of the true relationship based on information about the observed relationship and measurement quality:

$$r'_{AB} = \frac{r_{AB} - E^c_{AB}}{V_A V_B} \qquad (2)$$

A couple of examples will illustrate how the kinds of measurement quality estimates obtained in this study can be combined with information about an observed relationship to predict the true relationship between the underlying concepts.

1. Assume a correlation of .40 is observed between two measures that have estimated validities of .6 and .7 respectively, and that use different methods (hence we assume correlated error is zero). In this case, formula 2 predicts a true relationship of .95,

which is obviously much higher than the observed .40 relationship, which reflects the effects of random measurement error.

2. Assume a similar relationship, .40, is observed between two measures that have validities estimated at .93 and .95, respectively, and that use the same response scale and include method effects estimated at .36 (on the basis of which we assume correlated error is .13, which is .36 squared). In this case, formula 2 predicts a true relationship of .045, which is much lower than the observed relationship.

Multi-item scales

The estimates of validity and error components obtained in this study are for measures based on single survey items. Many survey analyses, however, use scales derived by combining several items. An important rationale for using such scales is that they usually have higher construct validity than single-item measures. (Depending on how a scale is constructed, it may also reflect higher method effects.) If one has an observed relationship involving one or more multi-item scales and one wishes to predict the underlying true relationship, the measurement quality of the scale(s) must be determined before formula 2 can be used.

In the simple and common situation where a set of items, all of which are assumed to tap the same underlying construct and have about equal validities and method effects, are added together to form a scale, a standard psychometric formula can be used to predict the validity of the scale. Guilford's (1954) formula 14.37 can be adapted for this purpose as follows:

$$V_s = \frac{V_I}{\sqrt{\frac{1 - V^2{}_I - M^2{}_I}{N + V^2{}_I + M^2{}_I}}} \qquad (3)$$

where
V_s is the estimated construct validity of the scale
V_I is the estimated construct validity of a single item
M_I is the estimated method effects in a single item
N is the number of items in the scale.

This same formula can be adapted to provide a prediction of the method effects reflected in a scale:

$$M_S = \frac{M_I}{\sqrt{\frac{1 - V^2{}_I - M^2{}_I}{N + V^2{}_I + M^2{}_I}}} \qquad (4)$$

where M_S is the estimated method effects in the scale, and all other terms are as above.

Once one has obtained estimates of the construct validity and method effects for a scale, the residual error effect can be obtained by the following formula:

$$V^R{}_S = \sqrt{1 - V^2{}_S - V^2{}_S} \qquad (5)$$

This is based on the definition that residual error is what is left after validity and method effects have been taken into account.

Predicting the validity and method effects for a simple addition of the items, if the items have different validities, and/or if the items reflect different method effects. There is some discussion of such situations in the literature (e.g. Green and Carmines, 1979), but the problem is complex and not fully solved. The new technology of structural modeling with latent variables, on which we comment below, offers a particularly useful way of handling some of these situations.

Multivariate relationships

Just as measurement errors affect bivariate relationships, so also do they have impacts on multivariate statistics, and here the effects are often harder to sort out. One approach for getting true multivariate relationships, uncontaminated by the effects of measurement error, is to use the procedures presented above for obtaining predictions of the true bivariate relationships, and then use these relationships as input to the calculation of the multivariate statistics. Another approach is to use structural modeling with latent variables, a powerful new approach for estimating the true relationships, either bivariate or multivariate, among underlying concepts (i.e., among latent variables). This is the structural modeling technology implemented in computer programs such as LISREL (Bentler, 1980; Jöreskog, 1978; Jöreskog and Sörbom, 1978).

If one had suitable data, one could use survey measures in such models and obtain useful estimates of underlying relationships without first making the corrections for measurement error just described. This requires, in effect, doing two things at once: estimating a set of measurement quality parameters and performing an analysis on the latent variables. However, sometimes one may not have data that permit simultaneous solution of both problems. In such circumstances, prior information about the validity and error components of the measures (information that might have been obtained using the approaches described in this report) can be incorporated into the structural equation model and allows one to proceed to an insightful analysis.

REFERENCES

Alwin, D. (1974). Approaches to the interpretation of relationships in the multitrait-multimethod matrix. In H.L. Costner (Ed.), *Sociological methodology 1973-1974*. San Francisco: Jossey-Bass.

Alwin, D.F., & Jackson, D.J. (1979). Measurement models for response errors in surveys: issues and applications. In K.F. Schuessler (Ed.), *Sociological methodology 1980*. San Francisco: Jossey-Bass.

American Psychological Association (1974). *Standards for educational and psychological tests*. Washington DC: American Psychological Association.

Andrews, F.M. (1979). Estimating the construct validity and correlated error components of the rated effectiveness measures. In F.M. Andrews (Ed.), *Scientific productivity*. Cambridge: Cambridge University Press/UNESCO.

Andrews, F.M. (1984). Construct validity and error components of survey measures. In C.F. Cannel and R.M. Groves (Eds.), *Health survey research methods, 4th Conference*. Rockville, Maryland: National Center for Health Services Research.

Andrews, F.M., & Crandall, R. (1976). The validity of measures of self-reported well-being. *Social Indicators Research, 3*, 1-19.

Andrews, F.M., & Herzog, A.R. (1982). *The quality of survey data as related to age of respondent*. Working paper #8032. Ann Arbor MI: Institute for Social Research.

Andrews, F.M., Morgan, J.N., Sonquist, J.A., & Klem, L. (1973). *Multiple classification analysis*. Ann Arbor, MI: Institute for Social Research.

Asher, H.B. (1974). Some consequences of measurement error in survey design. *American Journal of Political Science, 18*, 469-85.

Bailar, B.A. (1976). Some sources of error and their effect on census statistics. *Demography, 13*, 273-86.

Bentler, P.M. (1980). Significance tests and goodness of fit in the analysis of covariance structures. *Psychological Bulletin, 88,* 588-606.

Bielby, W.T., & Hauser, R.M. (1977). Response error in earnings functions for nonblack males. *Sociological Methods and Research, 6,* 241-80.

Bielby, W.T., Hauser, R.M., & Fetherman, D.L. (1977a). Response errors of black and nonblack males in models of the intergenerational transmission of socioeconomic status. *American Journal of Sociology, 82,* 242-88.

Bielby, W.T., Hauser, R.M., & Featherman, D.L. (1977b). Response errors of nonblack males in models of the stratification process. *Journal of the American Statistical Association, 72,* 723-75.

Bohrnstedt, G.W., & Borgatta, E.F. (1981). *Social measurement: Current issues.* Beverly Hills: Sage.

Bohrnstedt, G.W., & Carter, T.M. (1971). Robustness in regression analysis. In H.L. Costner (Ed.), *Sociological Methodology 1971.* San Francisco: Jossey-Bass.

Bollen, K.A., & Barb, K.H. (1981). Pearson's r and coarsely categorized measures. *American Sociological Review, 46,* 232-39.

Boruch, R.F., Larkin, J.D., Wolins, L., & MacKinney, A.C. (1970). Alternative methods of analysis: multitrait-multimethod data. *Educational and Psychological Measurement, 30,* 833-53

Campbell, D.T., & Fiske, D.W. (1959). Convergent and discriminant validation by the multitrait multimethod matrix. *Psychological Bulletin, 56,* 81-105.

Cochran, W.G. (1968). The effectiveness of adjustment by subclassifications in removing bias in observational studies. *Biometrics, 24,* 295-313.

Cochran, W.G. (1970). Some effects of errors of measurement on multiple correlation. *Journal of the American Statistical Association, 65,* 22-34.

Conner, R.J. (1972). Grouping for testing trends in categorical data. *Journal of the American Statistical Association, 67,* 601-04.

Corcoran, M. (1980). Sex differences in measurement error in status attainment models. *Sociological Methods and Research, 9,* 199-217.

Cox, E.P. (1980). The optimal number of response alternatives for a scale: a review. *Journal of Marketing Research, 17,* 407-22.

Cronbach, L.J., & Meehl, P.E. (1955). Construct validity in psychological tests. *Psychological Bulletin, 52,* 281-302.

Crowne, D.P., & Marlowe, D. (1964). *The approval motive: Studies in evaluative dependence.* New York: Wiley.

Fellegi, I.P. (1974). An improved method of estimating the correlated response variance. *Journal of the American Statistical Association, 69,* 496-501.

Green, V.L., & Carmines, E.G. (1979). Assessing the reliability of composites. In K.F. Schuessler (Ed.), *Sociological methodology.* San Francisco: Jossey-Bass.

Guilford, J.P. (1954). *Psychometric methods.* New York: McGraw-Hill.

Hansen, M.H., Hurwitz, W.N., & Bershad, M.A. (1961). Measurement errors in censuses and surveys. *Bulletin of the International Statistical Institute, 38,* 359-74.

Heise, D.R., & Bohrnstedt, G.W. (1970). Validity, invalidity, and reliability. In E.F. Borgatta and G.W. Bohrnstedt (Eds.), *Sociological methodology.* San Francisco: Jossey-Bass.

Hill, D.H. (1979). *A methodological note on obtaining MCA coefficients and standard errors.* Working Paper #8008. Ann Arbor, MI: Institute for Social Research.

Jöreskog, K.G. (1978). Structural analysis of covariance and correlation matrices. *Psychometrika, 43,* 443-77.

Jöreskog, K.G., & Sörbom, D. (1978). *LISREL IV users guide.* Chicago: National Educational Resources.

Kluegel, J.R., Singleton, R., Jr., & Starnes, C.E. (1977). Subjective class identification: A multiple indicator approach. *American Sociological Review, 42,* 599-611.

Krotki, K.P. (1978). *Estimation of correlated response variance.* Paper presented at the 1979 annual meeting of the American Statistical Association.

Krotki, K.P., & MacLeod, A. (1979). *Two methods of measuring correlated response variance.* Paper presented at the 1979 annual meeting of the American Statistical Association.

Linn, R.J., & Werts, C.E. (1973). Errors of inference due to errors of measurement. *Educational and Psychological Measurement, 33,* 531-544.

Lissitz, R.W., & Greene, S.B. (1975). Effect of the number of scale points on reliabilty: A Monte Carlo approach. *Journal of Applied Psychology, 60,* 10-13.

Mare, R.D., & Mason, W.M. (1980). A multiple group measurement model of children's report of parental socio-economic status. *Sociological Methods and Research, 9,* 178-198.

Martin, W.S. (1978). Effects of scaling on the correlation coefficient. *Journal of Marketing Research, 15,* 304-308.

Mason, W.M., Hauser, R.M., Kerckhoff, A.C., Poss, S.S., & Manton, K. (1976). Models of response error in student reports of parental socioeconomic characteristics. In W.H. Sewell et al. (Eds.), *Schooling and achievement in American Society*. New York: Academic Press.

Namboodiri, N.K., Carter, L.F., & Blalock, H.M., Jr. (1975). *Applied multivariate analysis and experimental design*. New York: McGraw-Hill.

Pearson, K. (1913). On the measurement of the influence of 'broad categories' on correlation. *Biometrica, 9,* 116-139.

Ramsey, J.O. (1973). The effect of number of categories in rating scales on precision of estimation of scale values. *Psychometrica, 38,* 513-32.

Robins, P.K., & West, R.W. (1977). Measurement errors in the estimation of home value. *Journal of the American Statistical Association, 72,* 290-294.

Schmitt, N., Coyle, B.W., & Saari, B.B. (1977). A review and critique of analyses of multitrait-multimethod matrices. *Multivariate Behavioural Research, 12,* 447-78.

Sonquist, J.A., Baker, E.L., & Morgan, J.N. (1974). *Searching for structure.* Ann Arbor, MI: Institute for Social Research.

Sudman, S., & Bradburn, N.L. (1974). *Response effects in surveys.* Chicago: Aldine.

Sullivan, J.L., & Feldman, S. (1979). *Multiple indicators: An introduction*. Beverly Hills: Sage.

Survey Research Center Computer Support Group (1981). *OSIRIS IV: statistical analysis and data management software system.* Ann Arbor, MI: Institute for Social Research.

Wheaton, B., Muthen, B., Alwin, D.F., & Summers, G.F. (1977). Assessing reliability and stability in panel models. In D.R. Heise (Ed.), *Sociological methodology 1977*. San Francisco: Jossey-Bass.

Zeller, R.A., & Carmines, E.G. (1980). *Measurement in the social sciences*. New York: Cambridge University Press.

chapter 2

Models for evaluation of measurement instruments

W.E. SARIS

Abstract

Recently there has been an increased interest in the quality of survey and panel data. Beside sampling errors, nonsampling errors are receiving particular attention. Campbell and Fiske (1959) suggested the use of different methods combined with different traits in order to evaluate measurement instruments. This has led to an evaluation called the multitrait multimethod approach, which has been used by Andrews (1984) to evaluate many different measurement instruments across studies.

In 1969 Heise suggested that test-retest studies could be the most appropriate method for the evaluation of survey questions in panel research. For this purpose he developed a model which was later improved by Wiley and Wiley (1970). The same model has recently been used by Alwin (1989) to evaluate measurement instruments. In accordance with the model used, it is known as the Quasi Simplex approach.

Jöreskog (1971) suggested that measurement instruments be evaluated with respect to validity by a specific congeneric test model. This model has been used by Saris (1982) for this purpose. This approach will be called the repeated multimethod approach. In this paper we discuss the possibilities of all three models for describing the qualities of measures of the same variable. On the basis of this discussion it is suggested that the repeated multimethod model approach be used for further research of the quality of measurement instruments.

INTRODUCTION

In this paper procedures for the evaluation of nonsampling errors in attitude research based on survey or panel data will be discussed. Recent studies by Andrews (1984), Alwin (1989) and Saris (1982) each represents a certain approach in the evaluation of measurement instruments. Andrews followed the multitrait multimethod approach originally introduced in the social sciences by Campbell and Fiske (1959), who suggested that the quality of measurement instruments can only be determined by comparing them with other instruments. This approach leads to special correlation matrices which are known as multitrait multimethod matrices. Andrews added to this approach by collecting such matrices for many topics, analyzing these data and then carrying out a meta-analysis of the obtained results in order to indicate the formulation aspects which affected the quality of the instruments.

Alwin (1989) used a different model, originally introduced in the social sciences by Heise (1969) and further developed by Wiley and Wiley (1970), which is known in the literature as the Quasi Simplex model (Jöreskog, 1971). Alwin added to this approach with a meta-analysis across many data sets in order to distinguish the effect of different formulation aspects on the qualities of different measurement instruments.

Both approaches have their advantages and disadvantages. In the following sections we will describe these and consequently suggest an alternative third model which, in a sense, is a combination of the two previous models. In its basic form this model was suggested by Jöreskog (1971) in a paper on the congeneric test model and has been applied by Saris (1982) for the kind of purposes discussed here.

An overview is given of the desired quality criteria for measurement instruments. The MTMM and the Quasi simplex approaches are then discussed with respect to obtaining this information. Finally we introduce a third approach which we believe leads to better results.

CHAPTER 2

QUALITY CRITERIA FOR MEASUREMENT INSTRUMENTS

The literature on validity and measurement error is rich and varied, but also imprecise and inconsistent (Andrews, 1984). Therefore it is necessary to say a few words about the different criteria that are used in this paper. Some of the most important issues in measurement are:
1. validity
2. invalidity
3. reliability

Books have been written about each of these concepts and we do not pretend to give the final answer here. An attempt will be made to provide a formal framework to indicate what we mean by the concepts used. Our approach is mainly in line with the formulation of Heise and Bohrnstedt (1970).

The discussion is restricted to the measurement of subjective variables such as attitudes, opinions, preferences, judgements, etc. None of these variables are directly observable. Therefore we normally use questions to obtain answers which are interpreted as indicators of the latent variables of interest. In this process we should always allow for random errors to occur. These errors may be due to misunderstandings, typing and coding errors, wrong interpretations of the question, etc. Representing the unobserved subjective variable of interest, by F and the response given to the question using method 1 by Rm_1 and the error by Em_1, we can formulate the simple model:

$$Rm_1 = F + Em_1 \qquad (1)$$

with
$cov(F,Em_1) = 0$
$E(Em_1) = 0$

Note that only the observed variable and the error variable have an index m_1 because F is assumed to be independent of the measurement procedure used. It would be very gratifying if all our measurement instruments satisfied this simple model. Unfortunately this is not the case. One deviation from the model specified occurs when the scales of the observed and latent variables are different. In model (1) it is assumed that the scales of the latent and observed variables are the same and that the mean of the one is the mean of the other. This is of course a somewhat arbitrary decision as we are discussing the scale of an unobserved variable. However, later we will see that we have to take into account the fact that the arbitrary chosen scale for the latent variable is not the same as the scales for all the observed variables. We therefore have to introduce scaling constants in equation (1) to correct for the differences between scales. Restricting ourselves to linear transformations, we get:

$$Rm_1 = am_1 + bm_1F + Em_1 \qquad (2)$$

with the same assumptions as before. The coefficients a and b are the scaling constants, which have an index because they can vary according to the method used. Model 2 should be seen as a more general formulation of the simple model of equation (1). Because of its generality, we shall refer to this model in the rest of this paper.

There have been many studies which suggest that responses are not only affected by the opinion of the respondent and some random error but also by some other potentially influential factors such as the opinion of the interviewer, the tendency to say yes or no, social desirability, etc. If these variables are present they will produce a systematic error in the responses, which is why we speak of invalidity in such cases. Normally these variables are also not directly observable. If we denote the sum of all these variables with Um_1 we can adjust equation (2) by introducing this new variable as an additive factor:

$$Rm_1 = am_1 + bm_1F + Um_1 + Em_1 \qquad (3)$$

where
$cov(F,Um_1)=0$
$cov(F, Em_11)=0$
$cov(Um_1, Em_1)=0$

It will be clear that the two disturbance terms Um_1 and Em_1 cannot be estimated separately under normal circumstances. Later we will show when separation of the two terms is possible.

A different formulation of the same model has also been used for the purpose of simplification. Equation (3) shows that the sum of the first three terms on the right hand side should remain the same if we repeat the same question under exactly the same conditions. Then only the last term, which is the random error component, would change. Therefore we could call the stable part, Tm_1, the true score for this measurement instrument and write:

$$Tm_1 = am_1 + bm_1F + Um_1 \quad (4.1)$$
and
$$Rm_1 = Tm_1 + Em_1 \quad (4.2)$$

with the same assumptions as mentioned above. Models (3) and (4) are of course compatible because substitution of the first equation in the second one of model (4) will lead to the equation of model (3). In this way the stable part of the response can be described separately from the random response process. This should not be taken too literally, as if there were two processes. It is only a way of writing in order to make the formulation and the definition simple.

A frequently cited definition of *reliability* is given by Lord and Novick (1968):

$$\text{reliability} = \frac{var(Tm_1)}{var(Rm_1)} \quad (5)$$

This is equal to the explained variance in Rm_1 or the squared standardized regression coefficient of equation (4.2), which could be formulated as:

$$R^sm_1 = \lambda^s{}_1T^sm_1 + Em_1 \quad (6)$$

where s denotes "standardized" , since

$$\lambda^s{}_1 = \sqrt{\frac{var(Tm_1)}{var(Rm_1)}}$$

it follows that the reliability = $(\lambda^s{}_1)^2$.

A similar formulation of *validity* and *invalidity* can be specified. If we standardize equation (4.1) we get:

$$T^sm_1 = a^sm_1 + b^sm_1F^s + Um'_1 \quad (7)$$

As cov(F,Um1)=0 it follows that the variance of T^sm_1, which is equal to 1, can be decomposed to the proportion of explained variance, which is $(bm^s{}_1)^2$, and the proportion of unexplained variance, which is equal to the variance of Um_1. Since F was the variable which we were supposed to measure, $(bm^s{}_1)^2$ is then the validity of the measurement procedure M_1 and $var(Um_1)$ the invalidity.

Even if one does not accept this definition of validity and invalidity, the given criteria for the quality of measurement instruments are important. In general, measurements with a high validity and reliability but a low invalidity are preferred. One advantage of the proposed formulation is that different criteria are defined independently. Other criteria can be derived from these concepts very easily as we will discuss later.

Reliability and validity, as they have been defined here, should be distinguished from each other because instrument characteristics can affect both of them and it would be useful to be able to distinguish these effects. For example, reduction of the number of categories in an instrument (reduction of precision) will increase its reliability because subjects can be expected to give approximately the same answers. However, the lack of precision will, under certain conditions, lead to large measure specific errors, as is shown in figure 1.

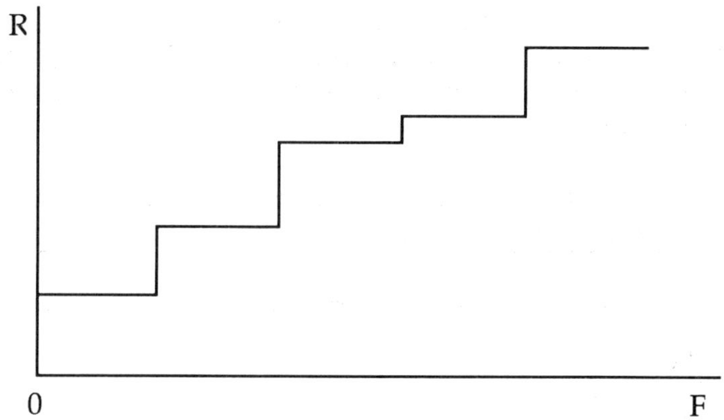

figure 1: A step function due to the crude categorical measurement of the attitude of interest

Another example of such a problem is the effect of variation in the way people express themselves, or variation in the response functions (Saris, 1988). This variation can exist across persons but also within persons from occasion to occasion. In order to determine which of the two effects is more prominent, a distinction must be made between reliability and invalidity.

The major problem addressed in this paper is how we can empirically determine the quality of instruments on the three criteria specified above. It should be clear from the start that information must be collected on more than one variable. This may be the same variable through time, or a comparison of different responses obtained with different instruments, or both. The discussion will be about the choice we will make in this respect.

It is impossible to distinguish reliability and validity if no repeated observation with the same variable is available as the two equations (4.1) and (4.2) will not be identified. With one observation, only one disturbance term, which will be a combination of Um_i and Em_i, can be estimated.

We shall now discuss three different approaches to estimating the quality indicators of an instrument. After this discussion we shall propose a specific procedure.

THE MULTITRAIT MULTIMETHOD APPROACH

Campbell and Fiske originally suggested the use of different methods for different traits in order to get an impression of the quality of the measurement instruments used. Since then many papers have been written about this approach. In the beginning the correlation matrix itself was interpreted. An example of such a matrix is given in table 1.

table 1: A multitrait multimethod matrix from a study of changes in the Dutch Trade Unions

	Rm_{11}	Rm_{21}	Rm_{31}	Rm_{12}	Rm_{22}	Rm_{32}	Rm_{13}	Rm_{23}	Rm_{33}
Rm_{11}	1.0								
Rm_{21}	.841	1.0							
Rm_{31}	.849	.906	1.0						
Rm_{12}	-.111	-.142	-.098	1.0					
Rm_{22}	-.053	.072	.038	.779	1.0				
Rm_{32}	-.071	-.011	.044	.811	.893	1.0			
Rm_{13}	.335	.327	.297	-.382	-.262	-.308	1.0		
Rm_{23}	.303	.433	.353	-.334	-.077	-.177	.778	1.0	
Rm_{33}	.336	.401	.402	-.376	-.197	-.192	.821	.893	1.0

Rm_{ij} is the response on a question asked using method i about topic j

This example with three traits and three methods is typical for this approach. Also, the sizes of the correlation coefficients do not differ very much from the normal case. The matrix itself is no longer interpreted and instead a structural equation model is specified to explain the correlations in a multitrait multimethod matrix.

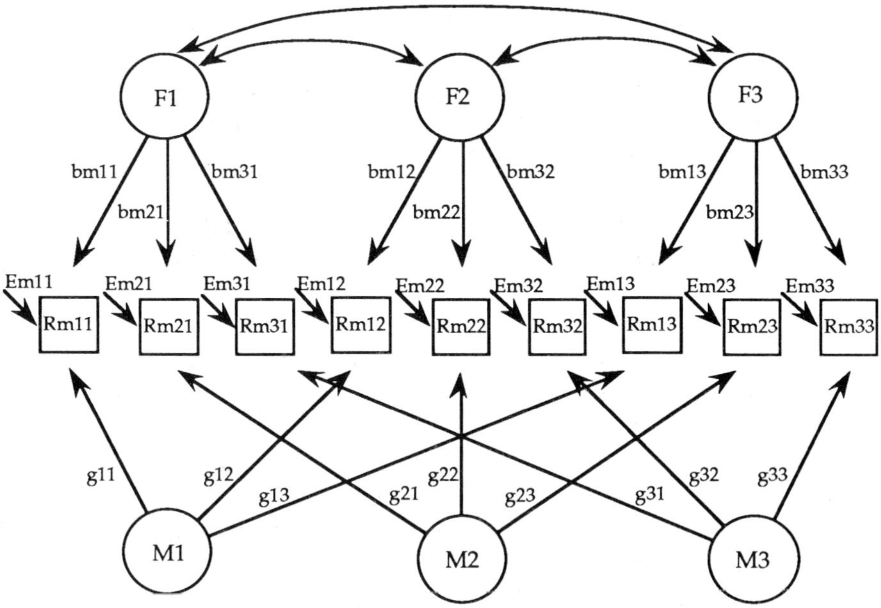

figure 2: An example of an MTMM model

Andrews (1984) used this approach in order to evaluate the instruments in his study. The model he used is comparable with that presented in figure 2.

This model can be formulated in the following equations:

$$Rm_{ij} = bm_{ij}F_j + g_{ij}M_i + Em_i \tag{8}$$

with

$cov(F_i,Em_j)=0, cov(M_i, Em_j)=0$ for all i,j

$cov(F_jM_i)=0$ for all i,j

$var(F_i)=var(G_i)=var(Rm_{ij})=1$ for all i,j

where Rm_{ij} is the response for the jth trait using the ith method. M_i is the variation which is due to the ith method. In itself this equation is similar to equation (3). The only difference is that the factor Um_1 in (3) is substituted for a specific factor M, which represents all the effects of the method on the responses. Although this formulation is elegant, it does have its limitations. The advantage of the design is that the method effects can be detected. However, other factors which are not related to the method will remain undetected as the estimation of the parameter g_{ij} is based on the correlation between the responses obtained with the same method.

Types of problems that can be detected are for example correlations which are due to crude category scales and those which are due to variation in response functions. However, both effects will only be picked up if, respectively, there is a certain stability in the way respondents place the cutting points in the continuous scales and/or in the way they answer the questions for different variables which are asked with the same method. For the latter kind of stability some evidence is available (Rule and Curtis, 1977; Saris, 1988).

Given that only a specific type of invalidating factors will be detected while others remain undetected, the error terms Em_i will contain all other factors which cause invalidity and/or unreliability. These two kinds of errors can therefore not be distinguished.

Besides these substantial problems, the specified MTMM design has many more technical problems. Firstly the solutions are very unstable: a typical solution has standardized bmi coefficients of .85 to .95 while the g_{ij} coefficients have estimated values of .2 to .3. The bmij coefficients are thus based on correlations around .6 to .8 while the method factors are based on correlations not larger than .04 or .09. The correlation between the method factors is also based on very small values. A correlation between the method factors of .75 produces only a correlation of .03 between the observed variables given the usual method effects. It will be clear that very large samples are required in order to get stable results for these method factors and the correlations between them.

These models are also plagued by a large number of unacceptable or nonconverging solutions. This is also a consequence of the small values of the correlations from which the parameters have to be estimated. For one factor it is easily possible that by sampling fluctuation two correlations between the observed variables will be positive and one negative. In such cases a factor model will have no solution.

A third problem is that there is a large series of models from which one can choose. The simplest model is that with only three trait factors, the largest is that with three trait factors and three method factors and correlations between the factors in each set but not between the factors from the different sets. Between these two extremes 16 different potentially applicable models can be specified (see figure 3).

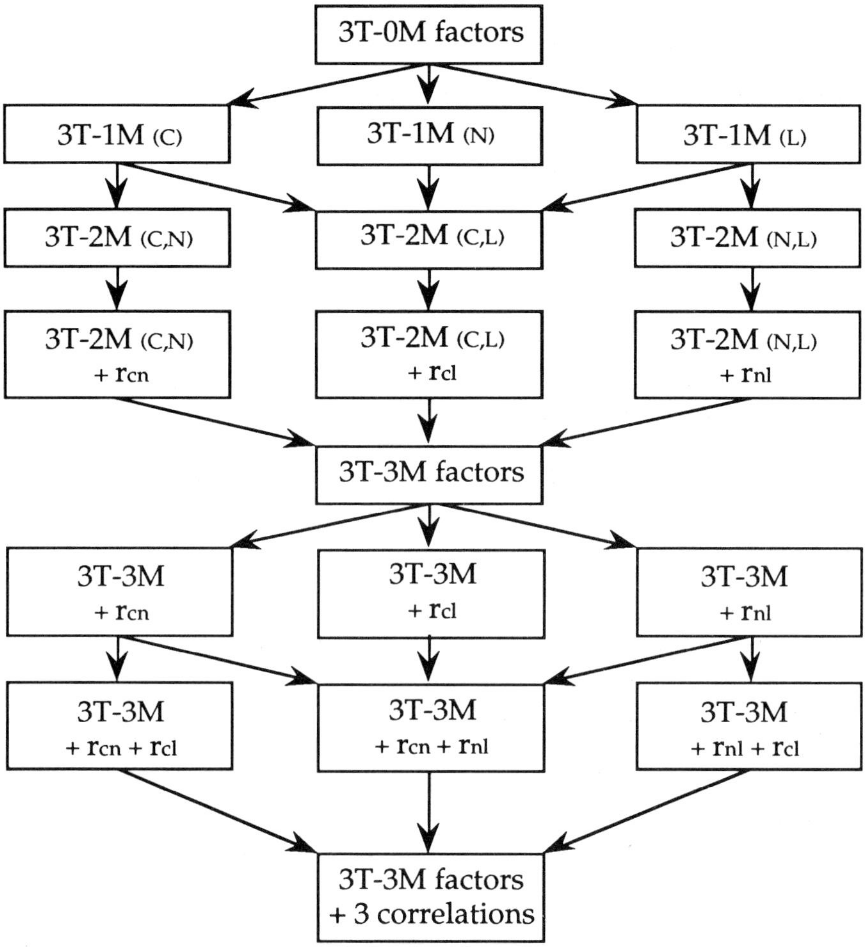

T = traits, M = method, C = category scale,
N = number scale, L = line scale

figure 3: A hierarchy of models for the analysis of MTMM matrices

It is not clear how one should choose between the different models. With the χ^2 test the decision depends very heavily on its power (Satorra and Saris, 1985), which has been shown to depend not only on the sample size but also on the size of the parameters (Saris and Satorra, 1988). Consequently more method factors may be found with better measurement of the traits (bm_{ij}).

table 2: Results from two population studies: with two loadings of .95 and one of .85 for each trait, and with two loadings of .85 and one of .75

data structure	fit of 3 factor model				
	ncp	df	largest resid	largest MI	largest EPC
bm_{ij} = .95/.85 (j=1-3) g_{ij} = .20 (j=4-6)	239	24	.04	61.5	-.638*
bm_{ij} = .85/.75 (j=1-3) g_{ij} = .20 (j=4-6)	16.4	24	.04	3.1	-.285*

* the largest EPC value is obtained for the wrong parameter

This can easily be demonstrated for a population study where models withthe same structure but different parameter values for the bmij parameters were applied. The data were generated with an MTMM model, but analyzed ignoring the method effects. Table 2 gives a summary of the results of this experiment.

The results show that the non-centrality parameter (ncp) or the power of the test is much higher with slightly higher values of the bmij parameters. In practice this would mean that with errors of the same size, the same model would be more frequently accepted when parameter values are lower. This is, of course, very unpractical.The use of fit criteria such as GFI, AGFI or the fit indices of Bentler are of no help because being based on the model test statistic they are also affected by the power of the test.

The modification index (Jöreskog and Sörbom, 1983) is affected in the same way by the power. The EPC measure suggested by Saris et al. (1987) should not have these problems but in this specific case these coefficients are also not very informative because they indicate that the parameters to be changed are very different from those which were wrongly specified. These bad results indicate that the fitted model was still very far from the correct model (Satorra et al., 1989). In this particular case it seemed that the fitted models were often very far from the true model even though the residuals were very small. The program Tetrad, which has been promoted recently for the detection of alternative models for the same data, was also unable to suggest suitable model corrections in this case (Glymour et al., 1987). This overview shows that our statistical tools are not sufficient for this type of model: perhaps, this problem is more general than we are willing to admit.

table 3: The results of an analysis with three different models of the MTMM matrix of table 1

	parameters	3F model	6F model	6F model (one corr factor)	
TRAIT EFFECTS					
category var.	λm_{11}	.886	.916	.913	
	λm_{12}	.844	.874	.927	
	λm_{13}	.847	.866	.879	
number var	λm_{21}	.949	.932	.904	
	λm_{22}	.927	.916	.864	
	λm_{23}	.919	.882	.845	
lines var	λm_{31}	.956	.946	.918	
	λm_{32}	.962	.943	.887	
	λm_{33}	.971	.964	.924	
METHOD EFFECTS					
categories	g_{11}	-	.361	.055	(ns)
	g_{12}	-	.157	-.135	(ns)
	g_{13}	-	.169	.329	(ns)
numbers	g_{21}	-	.239	.330	
	g_{22}	-	.299	.476	
	g_{23}	-	.276	.400	
lines	g_{31}	-	.144	.271	
	g_{32}	-	.224	.436	
	g_{33}	-	.212	.339	
corr factors		-	-	.68	
FIT OF THE MODELS					
	χ^2	200.1	100	20.6	
	df.	24	15	14	
	GFI	.799	.955	.99	
	AGFI	.624	.865	.97	
	RMR	.067	.058	.047	

In order to illustrate the problems of this approach we have performed an analysis with a three factor model on the correlation matrix given in table 1. This analysis gives a χ^2=200 with df=24. This model has to be rejected according to the usual criteria. With a six factor model there was also a bad fit to the data (χ^2=100 with df=15). If one correlation between the number and line factor is introduced the model fits (χ^2=20.6 and df=14). The corrections introduced in this model give a much better fit but, looking at the residuals, the improvement is minimal from one model to the next.

Even though the changes in the residuals are very small, the effect on the fit of the model is very large and the changes also have important effects on the values of the parameters, as can be seen in table 3.

This table shows that estimated values of the category variables were much lower (.06 to .13) than the loadings for the continuous variables in the three factor solution but that these differences decrease considerably in the 6 factor solution and even further if the correlation between the continuous factors is introduced. In the last case the category scale loadings are sometimes even larger. The sizes of these differences, which are the issue for evaluation studies, seem to be dependent on the model. Consequently, they are dependent on the power of the test and therefore on the size of the parameters themselves.

A further complication in these analyses is that the same respondents produced a fitting 5 factor solution for another topic with categories and numbers as method factors. If a 6 factor model was fitted the procedure did not converge. This illustrates that one can not simply use the least restricted model to avoid all the problems mentioned above.

The difference in results between the two topics can be explained by the fact that the respondents knew more about the second topic than about the first. When the sample was split into respondents who had some knowledge about the topic and those who had no knowledge at all, completely different correlation matrices were found. These matrices are presented in table 4.

The pooled data of these respondents are shown in table 1. Analysis of the matrix for the respondents who had no knowledge would of course lead to a completely different solution. It is in fact hard to find any fitting MTMM model for these data. It seems that respondents behave differently when they answer questions on topics which they have no knowledge about.

On the basis of the above arguments, I do not think that the design of the MTMM approach is the best for detecting the qualities of different measurement procedures. The major problems are:

- this design can detect the invalidity of method-connected factors but not other invalidating effects
- the estimates are very sensitive to minor differences in the parameter values and sample sizes due to the effect of these factors on the power of the test
- the test of the sequence of possible models is not clear

table 4: Two correlation matrices: for people with some knowledge of the topic (each first line) and for people with very little knowledge of the topic (each second line)

	Rm_{11}	Rm_{21}	Rm_{31}	Rm_{12}	Rm_{22}	Rm_{32}	Rm_{13}	Rm_{23}	Rm_{33}
Rm_{11}	1.0								
	1.0								
Rm_{21}	.89	1.0							
	.51	1.0							
Rm_{31}	.88	.93	1.0						
	.43	.74	1.0						
Rm_{12}	-.07	-.13	-.06	1.0					
	-.34	.02	-.13	1.0					
Rm_{22}	-.06	-.04	-.01	.88	1.0				
	.08	.68	.41	.22	1.0				
Rm_{32}	-.07	-.08	-.01	.89	.94	1.0			
	.04	.41	.53	.16	.68	1.0			
Rm_{13}	.29	.28	.28	-.39	-.34	-.36	1.0		
	-.02	.11	-.10	.68	.21	.05	1.0		
Rm_{23}	.23	.27	.24	-.40	-.34	-.34	.83	1.0	
	.09	.81	.59	.21	.74	.35	.24	1.0	
Rm_{33}	.27	.29	.31	-.41	-.36	-.32	.83	.91	1.0
	.04	.57	.70	.23	.37	.41	.31	.71	1.0

- there are many technical problems in the analysis. If this design is to be used, very detailed instructions should be formulated in order to prevent arbitrary differences between the different analyses.

But before we choose for this laborious operation, for which there is no obvious solution, we should look at other possible designs which serve the same purpose.

THE QUASI SIMPLEX APPROACH

An alternative to the MTMM approach is the quasi simplex approach. The model used here was first suggested by Heise (1969). He suggested that it is probably the most reasonable model to use for social science variables given that there are no batteries of questions which measure the same attitude or opinion. Therefore, the only way one can get an identified model in order to estimate the quality of an instrument is by repeated observations. He demonstrated that a design with three repeated observations is sufficient to get an identified model under mild restrictions. The basic model is presented in figure 4.

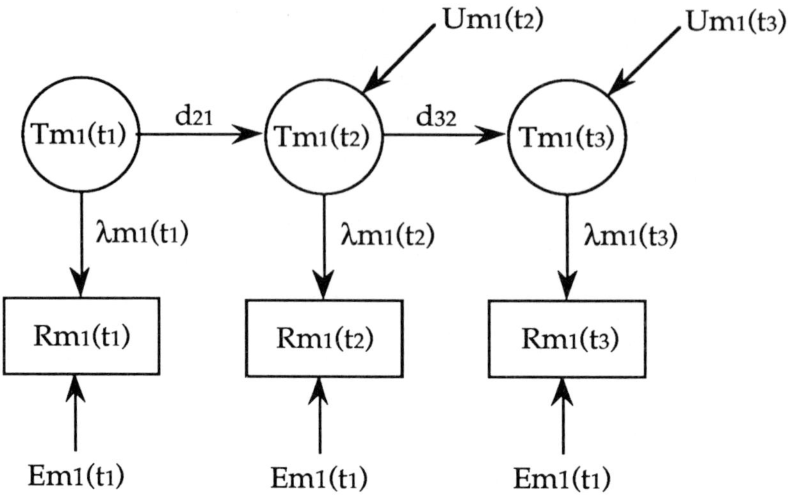

figure 4: The simplex model

Formally this model can be formulated as follows:

$$Rm_1(t_i) = \lambda m_1(t_i)\ Tm_1(t_i) + Em_1(t_i) \quad (9)$$
$$Tm_1(t_i) = d_{ij}Tm_1(t_i\text{-}1) + Um_1(t_i) \quad (10)$$

where

$cov(Tm_1(t_k), Em_1(t_i))=0$	for all i, k
$cov(Tm_1(t_k,Um_1(t_l))=0$	for all k≠l
$cov(Em(t_k),Um_1(t_l)=0$	for all k,l
$cov(Em(t_k),Em(t_{kl})=0$	for all k≠l
$cov(Um_1(t_k),Um_1(t_l)=0$	for all k≠l
$E(Tm_1(t_i) = E(Rm_1(t_i) = 0$	for all t_i

The first equation specifies the measurement model, which is in fact the same as equation (2) in the introduction. The second equation specifies a model for the possible changes through time of the true score (Tm_1).

This specification is not sufficient for identifying the parameters. For this purpose there are two different possibilities. The solution suggested by Heise was to assume that all variables are standardized and that the λm1 coefficients are identical for each point in time. It is therefore specified that the reliability is the same for each point in time, and the error variances will also be automatically identical through time. An alternative possibility is to work with unstandardized variables and to assume that $\lambda m_1=1$ at all points in time and that the error variances remain the same through time (Wiley and Wiley, 1970). Alternative models can also be formulated (Saris and van den Putte, 1988) but will not be discussed here. Alwin (1989) has chosen the formulation of Heise for his meta-analysis and tried to explain the variation in the size of the obtained parameter estimates from different data sets.

Although the measurement equation looks like equation (2), in which model no invalidity variables were introduced the data analyzed with the quasi simplex model could contain invalidating factors. However, the model specified does not and cannot take these factors into account. This would only be possible if more than one method has been used. Therefore we have to conclude that this model is not good at detecting invalidities in measurement procedures.

Nevertheless, the procedure is better than the MTMM model for determining the reliability of the variables. The standardized $\lambda m1$ coefficients provide a measure of the reliability of the measurement procedure. The only limitation of the specified model is that it must be assumed that the reliability or measurement error for all time points is the same if one has only three observations through time. When there are more than three observations, one can get an estimate of the reliability of the measurement procedure at any point in time, except the first and the last which remain unidentified (Werts, Jöreskog and Linn, 1971). Given the limited possibilities of this model for determining the invalidity and consequently also the validity of the instruments, we do not think that this model is appropriate for the evaluation of measurement instruments.

REPEATED MULTIMETHOD APPROACH

An alternative to the two previous not completely satisfactory designs is what we have called the Repeated Multimethod (RMM) approach. Similar approaches have been suggested by Jöreskog (1971) and by Saris (1982).

In this approach only one variable may be measured but at least two methods have to be used and the observations have to be made at least at two points in time with all measurement instruments. These two observations must be made within a short time interval so that it is reasonable to assume that the opinion or attitude has not changed. However, the time interval should be long enough for the observations to be seen as independent observations of the same variable.

If one trait is measured by three methods at two points in time, the model shown in figure 5 can be used to describe the data obtained.

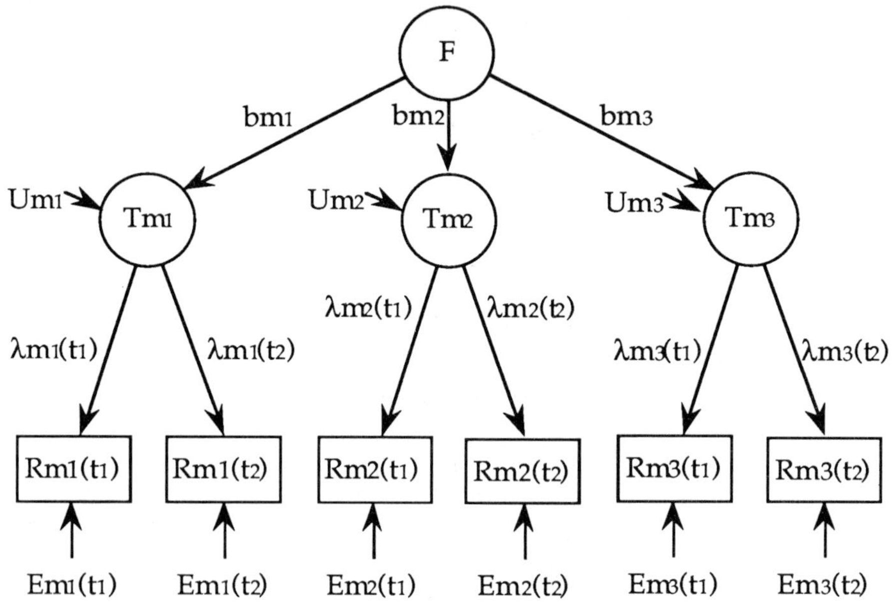

figure 5: The repeated multimethod model
for one latent variable

This model is completely in agreement with equation (4), which we have discussed above. $Rm_i(t_1)$ and $Rm_i(t_2)$ represent the two observations of the variable of interest with method i. Tm_i represents the systematic part or the true score for the observed responses obtained with the ith method. In order to obtain an identified model when only one latent variable is measured, at least three different methods have to be used. From a practical point of view (e.g. irritation of the respondents) this design is not the best. An alternative possibility is to measure two or more different traits. Only two different methods, which are used twice for each latent variable are then required. The two methods do not have to be the same for each trait. The model for this research design is presented in figure 6.

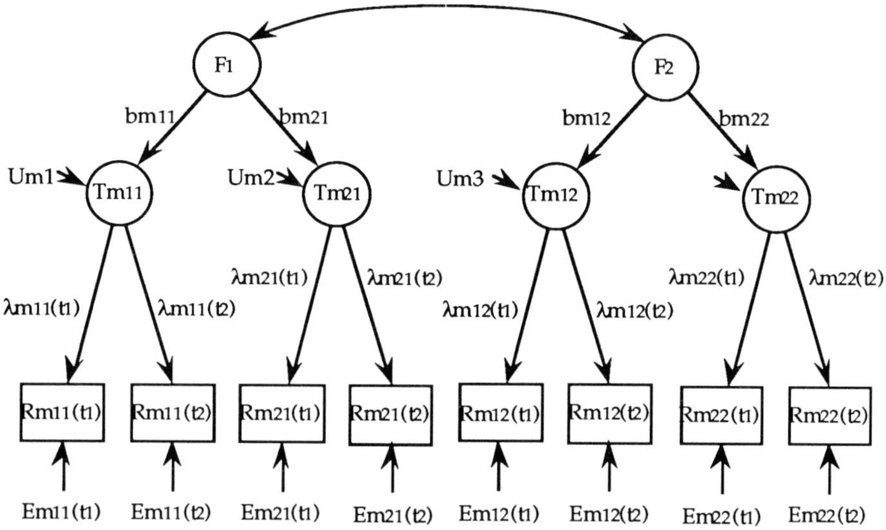

figure 6: The repeated multimethod model
for two observed traits

It can easily be seen that this model is again in agreement with equation (4). As there are now two traits, we have added an index to the variable which represents the traits. It can easily be verified that this model is also identified. It is even possible to test and estimate correlations between the unique variances in the true scores.

In these two models the *reliability* of the measurement instruments can be obtained from the squared standardized effect of Tm_i on Rm_{i1} or Rm_{i2}, or from the correlation between Rm_{i1} and Rm_{i2}, which should be the same if the two observed variables satisfy the parallel test requirements (Jöreskog, 1971). According to the previous definition, the *validity* is equal to the square of the standardized bmi coefficient for each method. The *invalidity* of each method is equal to the variance of the Um_1 terms.

The definition of the reliability is the same as for the quasi simplex model although the estimates can differ because of differences in the research design and sampling fluctuations. In the MTMM design the reliability cannot be distinguished from invalidity factors, which are not stable across topics. Therefore, with respect to the reliability estimates, the RMM has advantages above the MTMM design and is as good as the simplex approach.

With respect to the detection of invalidity, the RMM design is more attractive than the MTMM and simplex designs. The simplex design cannot detect the invalidities at all while the MTMM approach concentrates on method-specific factors which affect each observation when this method is used. If there is an interaction between a method and a variable of interest, i.e. an effect is specific for one variable of interest, then the specified MTMM model will not detect this invalidating effect (Browne, 1989). The RMM design can detect such factors. Therefore, the invalidity in this design will generally be at least as large as that in the MTMM design.

With respect to validity, the RMM design is better than the simplex approach which does not give this information. The choice between this and the MTMM approach is a matter of definition. Andrews (1984) has called the standardized effect of the factor F on the observed variable the validity coefficient. We would suggest the use of the squared standardized coefficients bm_{1j}, which indicate the effect of the factor on the systematic component. Andrews uses as his measure of validity the product of this coefficient with the reliability coefficient. The difference between these two approaches is simply a matter of taste. The advantage of the RMM approach is that the different criteria are defined independently and that the coefficient of Andrews can always be computed from the quantities obtained in the RMM approach. This is not the case if we start with the definitions of Andrews.

On the basis of the above discussion the RMM method seems more attractive than the other two approaches for comparative methodological research of measurement instruments. However, there are technical complications in the RMM approach. We have seen before that theoretical identification of a model does not always means that acceptable solutions are obtained.

Fortunately, the two models specified in figures 5 and 6 do not give many problems in practice. The estimates of the reliability coefficients λm_{ij} are usually quite high and the same is true for the validity coefficients bmij and consequently the estimation of these coefficients is also based on quite large correlations in the data. The invalidity coefficient in the equation is a residual of the true score and in general also leads to an acceptable solution, although sometimes negative values occur due to sampling fluctuation or omitted correlated errors somewhere in the model. The first possibility can be tested using the "t value" test on the residual variances as suggested by van Driel (1978). The latter problem can be avoided by careful experimental design, which is discussed in the last section.

The RMM approach also has the advantage that the number of models which could be fitted to the data is very limited. In a design with only one attitude to be evaluated and three methods the least restricted model would be the second order factor analysis model of figure 5. Alternative models can be formulated by introducing restrictions on the variances of the systematic errors (Um_1). Eventually a model with no systematic errors, which is in fact a one factor model, would be obtained. Although these restrictions could be introduced they complicate the testing, as has been mentioned above. It would be more efficient to estimate for all data sets the less restricted model (which does not lead to problems) so that an estimate of the invalidity is always obtained. If the data have been collected for two latent variables the model of figure 6 will apply and the same argument can be given. This model even includes the possibility of testing the stability of method-specific biasing factors.

Finally, there is the problem of the fit of the model. The models of figures 5 or 6 may not fit the data. There are at least two possible reasons for this. The first is that the power of the test is so high that the usual .05 level χ^2 test cannot be used. This situation is not unlikely since all coefficients, and therefore also the explained variances, are normally quite high. Under such circumstances the power of the test for small deviations from the model can be very high (Saris and Satorra, 1988) and the models will be rejected very frequently. In these cases, inspection of the residuals and the expected parameter change (EPC) might be useful. The residuals indicate how large the deviations of the estimated correlations are from the observed correlations. If they are smaller than an arbitrary chosen precision level (for example .05), then there is no need to worry about the fit of the model. Looking at the EPC's gives an idea of what would happen if parameters were added to the model. If the estimated values of these parameters are smaller than an arbitrary chosen level of precision, the model can be accepted even though the fit is not good.

There is, however also the possibility that the model does not satisfy these criteria. It is then necessary to reconsider the conditions under which the data have been collected. The most important reason why the model might be wrong is lack of independence of the observations. In most cases this is a consequence of a combination of factors: first a lack of information about the topic on the part of the respondents, and secondly a questionnaire design which allowed the respondents to answer the questions differently to how we expected them to. A typical example of such a situation is the battery of questions about the trade unions discussed above.

In that case many respondents had no knowledge and no opinion on the topic, but given the order of the questions, they produced some consistent data by remembering their previous answer and translating it in a different modality. Other examples of studies where this occurred have been reported by van Doorn et al. (1982) for questions about politics and by Seip et al. (1989) for psychophysical scaling experiments. The reasons for these correlated errors are dependent on the topic of study but should not be ignored. In the last section we will discuss different ways in which these experiments can be designed to avoid these dependencies as much as possible without giving the respondents reason to become irritated.

An example of the RMM approach

Before we discuss the research design for these analyses we first would like to give an example of a study where this design is used. In a nationwide telephone interview a random sample of 500 respondents was asked to answer questions on their voting behaviour:

1. *Are you going to vote in the next European election?*
   if not, how likely: maybe, probably not, certainly not
   if yes, how likely: probably, very likely, certain
2. *Can you indicate the probability that you are going to vote in a percentage: certainly not = 0 percent, certainly = 100 percent*

The same two questions were also asked for national elections and for local elections. For the same three types of elections we then asked:

how important is this election for you?

For each election we used one seven point category scale which was read to the respondents: extremely important, very important, important, do not know, unimportant, very unimportant, absolutely unimportant; and a specification of the responses on a numeric scale from 0 (absolutely unimportant) to 1000 (extremely important).

After these 12 questions a block of questions about preferences in national politics was presented. In this block probabilities were also asked. This block of questions took approximately 8 minutes to answer. After this block the interviewer had to read the following text aloud:

Since the first set of questions about the elections is very important for our study we would like to ask them again so that we are able to correct possible errors in the answers. Please do not think about what you said before but try to answer the questions as if they were new questions. This is also the fastest way to get through these questions.

The first twelve questions were then repeated. For our experiment we only used the first twelve questions which were repeated at the end. These twelve questions deal with 6 different traits: probability of voting in three different elections and the importance of these three elections. The methods used were the same for each election but different for the probability and the importance questions even though there was some similarity. For each topic we used a category and a numeric scale, but the answers on the category scale for the probabilities were obtained by a two-step procedure while the answers on the importance question were obtained by a usual seven-point scale. The numeric scales were also different. The probabilities were expressed in percentages from 0 to 100. For the importance questions the numeric scale ranged from 0 to 1000. These variations were introduced on purpose as we wanted the method correlation to be insignificant.

These data can be analysed in many different ways. For example, all traits can be analyzed at the same time, but then the model will be very large. The probability judgements and the importance judgements can be analyzed separately, but then the methods are the same for the different traits. For each election the probability judgements and the importance judgements could also be analyzed: the advantage of this model is that the methods for the two types of judgements are different, which might prevent method specific components from being correlated. We have chosen the last possibility for the analyses presented here. The first model used in this study is that of figure 6 with F1 representing the "probability to vote" and F2 the "importance of the election". Tm_{11} and Tm_{32} are the true scores for the categorical measurement procedures and Tm_{21} and Tm_{42} are the true scores for the numeric questions. The correlation matrix for the national election data is shown in table 5. In this table it can be seen that all correlations are rather high. Analysis of this matrix with the model of figure 6 gives the results presented in table 6.

table 5: The correlation matrix for data obtained using a RMM design with 2 different traits

	Rm_{11} (t_1)	Rm_{11} (t_2)	Rm_{21} (t_1)	Rm_{21} (t_2)	Rm_{12} (t_1)	Rm_{12} (t_2)	Rm_{22} (t_1)	Rm_{22} (t_2)
$Rm_{11}(t_1)$	1.000							
$Rm_{11}(t_2)$	.914	1.000						
$Rm_{21}(t_1)$	.937	.897	1.000					
$Rm_{21}(t_2)$	.897	.928	.923	1.000				
$Rm_{12}(t_1)$	.670	.688	.673	.700	1.000			
$Rm_{12}(t_2)$	.699	.737	.682	.734	.844	1.000		
$Rm_{22}(t_1)$	.673	.711	.696	.740	.855	.841	1.000	
$Rm_{22}(t_2)$	.674	.726	.703	.752	.801	.873	.905	1.000

table 6: The results of the analysis of the RMM matrix of table 5 with the model of figure 6

trait	method	time	reliability	validity	invalidity
1	1	1	.914	.983	.017
1	1	2	.914	.983	.017
1	2	1	.924	1.00	.000
1	2	2	.924	1.00	.000
2	1	1	.848	.976	.024
2	1	2	.848	.976	.024
2	2	1	.904	.953	.047
2	2	2	.904	.953	.047

Using the χ^2 test, we found that the model does not fit at the .05 level (χ^2 is 250 and df= 19). However, the loadings in this case are very high and therefore the power of the test is rather high, which means that the model will be rejected because of any small misspecification.

One possible misspecification is that the true scores have changed through time. A test which we performed showed that this possibility can be rejected. A second possibility is that the categorical response affected the continuous response, which followed immediately. This hypothesis could be tested and did not lead to a significant improvement of the model: the difference in χ^2=3 with df=4.

A third possibility is that the respondent makes a similar kind of error in both measures at each moment. This hypothesis can be tested by introducing correlated errors for the responses which are given at the same point in time for each topic. The introduction of these four covariances between the error terms led to a considerable reduction of the χ^2 but the model still did not fit (χ^2= 59.8, df=15).

At this point we corrected other parts of the model. An obvious possibility was the introduction of a method specific correlation. Introduction of a correlation between the U terms of two methods led to a reduction of χ^2 to 32.5 with 14 df. This is of course again a better fit but one might wonder what one is doing. The introduced covariance term is equal to .033. Although this coefficient is significant, it is mainly due to the high explained variance in this model that this small parameter has to be introduced. The effect of this extra parameter on the rest of the model is also extremely small. The reliability coefficients remained the same but the validity coefficients changed maximally by .013. This occurred for only one validity coefficient. For all other parameters the change was smaller than .01. A problem with this correction of the model is that there are two choices: the correlation between Um_{11} and Um_{12} or between Um_{12} and Um_{22}. Depending on the choice, the change is in a different direction. This suggests that both corrections should be introduced at the same time in order to estimate their effects. However, this is impossible because of identification reasons. Given the very small misspecification and its small effect, we decided not to introduce this correction in the model.

Exactly the same results have been obtained for the other two topics, where the correlations between the method specific components were even smaller (.019 and .022) respectively. Therefore, the same arguments apply in all three cases: the correlated errors are introduced but the correlations between the method specific components are ignored. The results of these analyses are summarized in table 7.

Although this study was carried out mainly for methodological reasons, the results are rather interesting. Table 7 shows that for the first topic (voting probability) the reliability and the validity are a little bit higher if continuous scales are used instead of categorical scales for each of the three topics whereas for the second topic (importance of the election) the reliability of the continuous scale is much higher than that of the categorical scale (4.6% to 6.2%). With respect to validity the situation is unclear: for one topic

table 7: The results of the analysis of the RMM matrix of table 5 with the model of figure 6 extended to include correlated errors

trait	method	time	reliability EC	reliability NAT	reliability LOC	validity EC	validity NAT	validity LOC
1	1	1	.914	.935	.960	.964	.966	.982
1	1	2	.914	.935	.960	.964	.966	.982
1	2	1	.918	.945	.970	.992	.988	.992
1	2	2	.918	.945	.970	.992	.988	.992
2	1	1	.854	.852	.906	.901	.949	.939
2	1	2	.854	.852	.906	.901	.949	.939
2	2	1	.916	.908	.951	.880	.933	.953
2	2	2	.916	.908	.951	.880	.933	.953

the continuous scales were slightly better, for the other two the category scales were better.

A total evaluation can also be made by multiplying the validity and reliability coefficients to give an indication of the strength of the relationship between the variable of interest (F) and the observed variables. The results are presented in table 8.

table 8: The total evaluation of the measurement instruments

trait	method	time	total evaluation (rel*val) EC	NAT	LOC
1	1	1	.881	.903	.943
1	1	2	.881	.903	.943
1	2	1	.911	.934	.962
1	2	2	.911	.934	.962
2	1	1	.770	.809	.849
2	1	2	.770	.809	.849
2	2	1	.806	.847	.906
2	2	2	.806	.847	.906

It is interesting to note that in all cases the continuous scales are at least 3% better than the categorical scales. It is also clear that in all cases the measures of voting probabilities are approximately 10% better than the measures of perceived importance. The coefficient of total evaluation is the squared validity coefficient used by Andrews (1984). The advantage of the RMM approach is that we can also decompose this coefficient to determine whether the better performance is due to the higher validity or the higher reliability.

In the MTMM approach we can also determine the method effect as Andrews has done. In this study there should be no method effects at all according to this approach. As the methods were not the same, no method effect could occur. In the RMM approach we can detect method specific effects, which we have called invalidity, by subtracting the validity from 1 (see table 6). This invalidity represents all specific components of the method.

It should be noted that no correlations between the unique components have been specified. In the MTMM approach this is the same as ignoring correlations between the method factors. We have seen above that this simplification leads to very minor specification errors if the design is properly chosen.

CHOICE OF A RESEARCH DESIGN

The MTMM and the RMM designs have the serious disadvantage that both require from the respondents a large number of repeated judgements about the same topic. In the MTMM approach the respondents will in principle have to answer the same question at least three times (with three different methods), and this should be done for at least three topics. In the RMM design, for one topic respondents must twice answer the same question asked in three different ways. For two topics they must twice give answers in at least two different ways.

Obviously, the respondents cannot be overloaded with such tasks, which have to be as easy as possible. By placing the questions in a logical sequence for the respondents, the tasks are made simpler. In practice there are two different ways one can order these questions:

- sequence of topics per method, repeated later
- sequence of methods per topic, repeated later.

The time interval between the same two questions should at least be long enough for there to be a chance that the respondents will have forgotten their previous answers. The choice between these two depends on the method used and the number of topics asked. From the respondent's point of view, the "method per topic sequence" is sometimes preferable, like in our example. This can easily be shown. For example it is very natural to ask:

How would you rate your house?
1. very good
2. good
3. neither good nor bad
4. bad
5. very bad
Can you also evaluate your house on a scale from 0- 1000
where 0 = the worst you can imagine and 1000 = an ideal house
(the larger the number the better your house)

After this set of questions, while varying the method but keeping the topic the same, other topics can be asked in the same way. It is possible that effects across methods will lead to more consistency across methods than actually exists. Van Doorn (1983) found no evidence for this when he compared the two approaches. The correlated error patterns in this study were not due to order effects but were related to substantive aspects of the questions.

This sequence is, however, not always possible. For example, if the formulation of a question is varied, the two forms cannot be asked in sequence. Therefore we must use the second possibility, even though it is less attractive because the respondent will have the impression that the same task has to be carried out four times. In the previous alternative the measurement with the two methods is seen as one approach which is later repeated for a limited number of topics.

This second possibility is also not free of sequential effects, as has been shown by Seip et al. (1989). If the items to be evaluated are very similar, it seems that the respondents are even inclined to use a different model to formulate their responses. With this approach we must therefore avoid orderings bu which similar items are very close to each other.

CONCLUSIONS

In summary, we have attempted to show that the RMM design should be chosen for comparative research. The major problems of the MTMM approach are:
- it can only detect the invalidity of method-related factors and will not detect other invalidating effects
- the estimates are very sensitive to minor differences in the parameter values and sample sizes
- the test of the sequence of possible models is not clear

- there are many technical problems in the analysis.

We have shown in our example that the problems involved using the RMM approach are much less and the information provided is more detailed. We also recommend the use of two or more different traits and two methods.

The sequence of questions to be chosen depends mainly on the topic and the variation introduced but, if possible, the variation in method per topic, with a repetition at a later point in time during the same interview for a few topics of the set, is recommended.

Finally, in our experiment we have tried to minimize the correlation between the method-specific components by using different methods for different topics.

REFERENCES

Alwin, D.F. (1989). Problems in the estimation and interpretation of the reliability of survey data. *Quality and Quantity, 23,* 277-333.

Andrews, F.M. (1984). Construct validity and error components of survey measures: A structural modelling approach. *Public Opinion Quarterly, 48,* 409-442.

Browne, M.W. (1984). The decomposition of multitrait multimethod data. *Educational and Psychological Measurement, 30,* 833-853.

Campbell, D.I., & Fiske, D.W. (1959). Convergent and discriminant validation by the multitrait-multimethod matrix. *Psychological Bulletin, 56,* 81-105.

Doorn, L. van, Saris, W.E., & Lodge, M. (1982). The measurement of issue variables, position of respondents, candidates and parties. In W.E. Saris (Ed.), *Sociometric research.* Amsterdam: Sociometric Research Foundation.

Driel, O.P. van (1978). On various cases of improper solutions in maximum likelihood factor analysis. *Psychometrica, 43,* 225-243.

Glymour, C., Scheines, R., Spirtes, P., & Kelly, K. (1987). *Discovery causal structure.* Orlando FL: Academic Press.

Heise, D.R. (1969). Separating reliability and stability in test-retest-correlation. *American Sociological Review, 34,* 93-101.

Heise, D.R., & Bohrnstedt, G.W. (1970). Validity, invalidity and reliability. In E.F. Borgatta and G.W. Bohrnstedt (Eds.), *Sociological methodology 1970.* San Francisco: Jossey-Bass.

Jöreskog, K.G. (1971). Statistical analysis of sets of congeneric tests. *Psychometrica, 36,* 109-133.

Jöreskog, K.G., & Sörbom, D. (1983). *LISREL VI users guide.* University of Uppsala: Department of Statistics.

Lord, F.M., & Novick, M.R. (1968). *Statistical theories of mental test scores.* Reading, MA: Addison-Wesley.

Rule, S.J., & Curtis, D.W. (1977). Subject differences in input and output transformations from magnitude estimation of differences. *Acta Psychologica, 41,* 61-65.

Saris, W.E. (1982). Different questions, different variables. In C. Fornell (Ed.), *A second generation of multivariate analysis: Vol. 2. Measurement and evaluation.* New York: Praeger.

Saris, W.E. (1988). *Variation in response functions: A source of measurement error in attitude research.* Amsterdam: Sociometric Research Foundation.

Saris, W.E., & Satorra, A. (1988). Characteristics of structural equation models which affect the power of the likelihood ratio test. In W.E Saris & J.N. Galhofer (Eds.), *Sociometric Research: Vol. 2. Data analysis.* London: MacMillan.

Saris, W.E., & Putte, B. van den (1988). True scores or factor models: A secondary analysis of the ALLBUS-test-retest data. *Sociological Method and Research, 17,* 123-157.

Saris, W.E., Satorra, A., & Sorböm, D. (1987). The detection and correction of specification errors in structural equation models. In C.C. Clogg (Ed.), *Sociological methodology 1987.* Washington DC: American Sociological Association.

Satorra, A., & Saris, W.E. (1985). Power of the likelihood ratio test in covariance structure analysis. *Psychometrica, 50,* 83-90.

Satorra, A., Saris, W.E., & Pijper, W.M. de (1985). *Several approximations to the power function for the likelihood ratio test in covariance structure analysis.* Amsterdam: Sociometric Research Foundation.

Seip, H., Maas, K., & Saris, W.E. (1989). Variation in response models for psychophysical scaling. In B. van den Putte (Ed.), *Individual and subgroup analysis: Five substantive examples.* Amsterdam: Sociometric Research Foundation.

Werts, C.E., Jöreskog, K.G., & Linn, R.L. (1971). Comment on the estimation of measurement error in panel data. *American Sociological Review, 336,* 110-113.

Wiley, D.E., & Wiley, J.A. (1970). The estimation of measurement error in panel data. *American Sociological Review, 35,* 112-117.

chapter 3

Analysis of multitrait multimethod matrices

R. KÖLTRINGER

A review of the psychological literature by Schmitt and Stults (1986) revealed that the following procedures are often used to analyze multitrait multimethod (MTMM) matrices:
- exploratory factor analysis
- smallest space analysis (SSA)
- analysis of variance, including
  a. nonparametric alternatives
  b. confirmatory factor analysis (CFA)

Of all these analytical strategies, only the CFA approach leads to numerical estimates of validity, method, and residual variance components of the indicators. Therefore the other procedures will not be described in detail.

THE GENERAL CFA MODEL FOR MTMM-DATA

Consider n persons who are rated using m different methods of measurement on t traits. This research design results in a (mt x 1) vector of observed variables **x**, which can be seen as a linear function of t + m factors ξ as follows:

$$\mathbf{x} = \Lambda \, \xi + \delta \tag{1}$$

where
- Λ is a (mt x (t + m)) matrix of factor loadings
- ξ is a ((t + m) x 1) vector of t trait and m method factors
- δ is a (mt x 1) vector of residuals.

Assuming that the residuals are uncorrelated with each other and with the factors, equation (1) gives the standard factor analytic decomposition

$$\Sigma = \Lambda \, \Phi \, \Lambda' + \Theta_\delta \tag{2}$$

where

- Σ is an (mt xmt) observed covariance (or correlation) matrix
- Λ is defined as in (1)
- Φ is a (t + m) x (t + m) covariance (or correlation) matrix of the t + m factors
- Θ_δ is an (mt x mt) diagonal matrix of residual variances.

In order to simplify further discussion it is useful to partition Φ as

$$\Phi = \begin{bmatrix} \Phi_{tt} & \Phi_{tm} \\ \Phi_{mt} & \Phi_{mm} \end{bmatrix}$$

where

- Φ_{tt} is a (t x t) symmetric submatrix of Φ that contains trait factor variances and covariance
- Φ_{mm} is an (m x m) symmetric submatrix of Φ that contains method factor variances and covariances
- Φ_{mt} (= Φ'_{tm}) is an (m x t) rectangular submatrix of Φ that contains covariances of the m method factors with the t trait factors

and Λ as

$$\Lambda = [\Lambda_t \,|\, \Lambda_m]$$

where

- Λ_t is an (mt x t) submatrix of Λ that contains zeros, except for the loadings of observed variables on their respective trait factors
- Λ_t is an (mt x m) submatrix of Λ that contains zeros, except for the loadings of the observed variables on their respective method factors.

Assuming that the factor analytic model of Σ represented by equation (2) is valid and that scores on the mt observed measures follow a multivariate normal distribution, Jöreskog (1969, 1971) developed procedures for obtaining maximum likelihood estimates of all model parameters in Λ, Φ, and Θ_δ. The significance of each of the parameters in the model may be tested by forming a z-value of the parameter estimate by dividing its asymptotic standard error. In addition to the test of each individual parameter estimate, maximum likelihood estimation yields an overall χ^2 goodness-of-fit test, which is a test of the difference in fit between a given model and a completely saturated model that perfectly reproduces the data.

Problems of the CFA-approach

Besides the difficulties inherent in covariance structure analysis as a whole (e.g. the rather restrictive distributional assumptions), CFA models for MTMM data have several additional problems. The full CFA model for MTMM data, as represented in equation (2), is only identified for $t \geq 3$ and $m \geq 3$. However, as Kenny (1976) pointed out, the model can, with further restrictive assumptions, be used to test MTMM matrices with smaller dimensions. It can be shown that the use of other methods applicable to smaller matrices requires the same (restrictive) assumptions (Alwin, 1974).

Even under the condition $t \geq 3$ and $m \geq 3$, trait-method covariances are empirically underidentified. Applications in which parameters in Φ_{mt} were estimated resulted in

- factor loadings or factor intercorrelations greater then 1.00 for standardized data,
- the presence of Heywood Cases (estimates of unique variance close to zero), and
- very large standard errors for some parameters (because of high intercorrelations of the parameter estimates).

Constraining Φ_{mt} to be a null matrix leads to orthogonality of the trait and method factor space, a condition that is highly desirable since it results in a decomposition of the variance of each indicator into an additive combination of trait, method, and residual (random error) variance. Our present knowledge about the empirical nature or practical importance of trait-method interactions is very limited. It is possible that small trait-method correlations do not effect the trait and method factor loading estimates substantively but exaggerate the residual variances. In order to provide guide-lines for the interpretation of CFA models for MTMM-data, simulation studies on this point are required.

When $t \geq 3$ and $m \geq 3$ and all the parameters in Φ_{mt} are fixed to zero, the CFA model for MTMM data may still be empirically underidentified because of overfactoring (Rindskopf, 1984). It is well known that at least three measures of each factor are needed to identify orthogonal factor models. Therefore, in the case of a CFA-model with a $t = 3$ and $m = 3$ MTMM-design, Heywood cases and related problems are very likely if some factor intercorrelations and factor loadings of the same factors are close to zero.

To ensure nonnegativity of unique variances in Θ_δ a procedure outlined by Rindskopf (1983) can be used: The technique requires the observed measures to be treated as y-variables and the specification of exogenous latent variables (ξ), whose variance is fixed to one, as trait, method, and residual variables. Instead of the usual parameterization of MTMM models, the factor loadings and then the residual effects are estimated in Γ, while the matrices Θ_δ, Θ_ε, and Φ are not used. Since it is not the unique variances that are directly estimated but the effects of the unique factors, the residual variances, which are the squares of these effects, will always be positive. However, it is clear, that an empirically underidentified model with some large standard errors and highly correlated parameter estimates will still be underidentified if the described parameterization is used.

Some researchers have suggested the inclusion of a general factor in the model represented in equation (2). This general factor can - in the presence of both trait and method factors - account for dimensions which inflate all the correlations in a given MTMM matrix (e.g. a general trait dimension or a response set factor). But the interpretation of such a general factor is rather indeterminate. It therefore seems to be preferable not to allow for this specification.

Representing the trends present in MTMM data is a complicated undertaking with many subjective components. Several authors who analyzed the same published MTMM matrix arrived at directly contradictory conclusions (Widaman, 1985).

HIERARCHICALLY NESTED MODELS FOR MTMM DATA

Most of the problems mentioned above can be circumvented by the strategy of nested model testing. Specifically, the probability of accepting a wrong model with a good fit and the danger of overfactoring may be minimized by the systematic comparison of various alternative models. In order to reduce the intuitive dimension of the stepwise testing procedure, Widaman (1985) proposed an array of models for MTMM matrices, based on three possible structures of trait and method factors:

- no trait (or method) factors;
- trait (or method) factors with fixed intercorrelations: either 0, indicating a high level of discriminant validity, or 1, indicating total lack of discriminant validity, and
- trait (or method) factors with freely estimated intercorrelations.

Cross-classifying of the two structures results in the taxonomy of models for MTMM data presented in table 1.

table 1:
Taxonomy of structural models for MTMM-data
(following Widaman 1985)

TRAIT STRUCTURE	METHOD STRUCTURE			
	no method factors	m method factors (fixed unit correlations)	m method factors (fixed zero intercorrelations)	m method factors (freely estimated intercorrelations)
no trait factors	**H0**: null model	1 general method (not clearly interpretable)	m methods only (orthogonal)	**A**: m methods only (oblique)
t trait factors (fixed unit inter-correlations)	1 general trait (not clearly interpretable)	2 general factors (not identified)	1 general trait + m methods (orthogonal)	**B**: 1 general trait + m methods (oblique)
t trait factors (fixed zero inter-correlations)	t traits only (orthogonal)	t traits (orthogonal) + 1 general method	t traits (orthogonal) + m methods (orthogonal)	t traits (orthogonal) + m methods (oblique)
t trait factors (freely estimated intercorrelations)	**H1**: t traits only (oblique)	t traits (oblique) + 1 general method	**H2**: t traits (oblique) + m methods (orthogonal)	**H3**: full model t traits (oblique) + m methods (oblique)

Note that table 1 is demarcated into nine areas. A particular model is nested within all other models separated by at least one line of demarcation lying to the right and/or below the model in question.

Fitting and comparing all of the models listed in table 1 might lead to the selection of a best-fitting model. But following this strategy is rather cumbersome. Widaman (1985) preferred another rationale, which is based on parsimony. With this approach it is reasonable to start with a highly restricted (but theoretically possible) model and relax restrictions only if a model fails to represent adequately a set of data:

H_0 Null Model:
Hypothesis: lack of correlation among observed variables.
Parameters: θ_δ (Λ_t, Λ_m, and ϕ_{mt} fixed to zero; $\phi_{tt} = \mathbf{I}$, $\phi_{mm} = \mathbf{I}$).

H_1 First alternative Model:
Hypothesis: covariation among observed variables is due only to trait factors and their intercorrelations.
Parameters: Λ_t,ϕ_{tt}, and θ_δ; (Λ_m, and ϕ_{mt} fixed to zero, $\phi_{mm} = \mathbf{I}$).

H_2 Second alternative Model:
Hypothesis: (orthogonal) method factors are also present).
Parameters: Λ_t,ϕ_{tt}, Λ_m, and θ_δ (ϕ_{mt} fixed to zero, $\phi_{mm} = \mathbf{I}$).

H_3 Full Model:
Hypothesis: (oblique) trait factors and (oblique) method factors account for the observed correlations.
Parameters: Λ_t,ϕ_{tt}, Λ_m, ϕ_{mm}, and θ_δ (ϕ_{mt} fixed to zero).

There is at least one problem concerning the above test strategy: the importance of trait factors for the reproduction of the MTMM matrix is tested against a highly restrictive Null Model (H_0). It can be argued that the probability of accepting the alternative model (significant trait factors) is high, even if the alternative model is wrong.

Nested model comparisons

In order to test the degree of convergent validity (and, if necessary the degree of discriminant validity) models A and B in table 1 must also be estimated. Assume that the Full Model H_3 represents the data quite well on statistical and practical grounds. Then, if the fit of H_3 (t oblique trait factors and m oblique method factors) is substantially better than the fit of model A (no trait factors, m oblique method factors), there is a significant amount of covariance among measures uniquely explained by the trait factors - this represents *convergent validity.*

To test for *discriminant validity,* the fit of model H_3 should be compared with the fit of Model B, which has fixed perfect intercorrelations among trait factors (and m oblique method factors). Comparison of the fit of the models H1 (oblique trait factors only) and H_2 (oblique trait and orthogonal method factors) gives a test statistic for the significance of the *method variance* among measures. Finally, the *discriminability of method effects* can be tested by comparing the models H_2 (t oblique traits and m orthogonal method factors) and H_3 (t oblique traits and m oblique method factors).

Table 2 summarizes the described nested model comparisons for MTMM matrices. As can be seen from this table, the coefficient Q (Jöreskog, 1974) is preferred as the a test statistic for the model comparisons (other possible measures are described in Loehlin, 1987, pp. 69-71). Q is defined as the ratio of decline in χ^2 to change in degrees of freedom; $Q > 2.0$ indicates that the improvement of fit is substantial and not merely the result of chance:

$$Q = \frac{\chi^2(H_x) - \chi^2(H_y)}{df(H_x) - df(H_y)}$$

Since subsequent models in the test series do not always differ by the same degrees of freedom, the PRE-coefficient Delta, proposed by Bentler and Bonett (1980) should not be applied. But this measure is very useful to assess the level of practical fit of a given model in addition to its statistical fit.

$$\text{Delta} = \frac{\chi^2(H_o) - \chi^2(H_x)}{\chi^2(H_o)}$$

The results of the model comparisons summarized in table 2 are "ideal" in the following sense: first, the proportion of covariation among observed measures uniquely representable as convergent validity (e.g. due to trait factors) is significant. Second, there is a pronounced degree of discriminant validity. Third, the covariation explained by method factors is also quite high.

Anomalies in the proposed test sequence most often occur when unnecessary factors are included in the model. As Rindskopf (1984) pointed out, overfactoring is present when a factor has

1. no large loadings, or

table 2: Nested model comparisons for MTMM data

MODELS

M_0	specifications	χ^2	df [1]	P	anomalies [2]
H_0	no trait factors no method factors	....	36	....	....
H_1	t traits only (oblique)	....	24	....	....
H_2	t (oblique) traits, m (orthogonal) methods	....	15	....	....
H_3	t (oblique) traits, m (oblique) methods	....	12	....	....
A	m methods only (oblique)	....	24	....	....
B	1 general trait, m (oblique) methods	....	15	....	....

MODEL COMPARISONS (predicted outcomes for "normal" MTMM matrices)

comparison testing for:			Q	accepted model [3]
H_0	H_1	significance of trait factors	....	H_1
H_1	H_2	significance of method factors	....	H_2
H_2	H_3	discriminability of method effects	....	H_3
A	H_3	convergent validity	....	H_3
B	H_3	discriminant validity	....	H_3

[1] For t=3 and m=3.

[2] Space for indicating the occurrence of Heywood Cases etc.

[3] $Q > 2$

2. only one large loading, or
3. only two large loadings, and close to zero correlations with all other so-called real factors (factors with two or more large loadings)

In all of these situations, the model will be empirically underidentified:

- In the first case, the factor can have correlations with other factors over a fairly wide range without substantially changing the reproduced covariance matrix.
- In the second case, the nonzero loading cannot be separated reliable from residual variance, and the result can be an impossibly large loading accompanied by a negative unique variance.

- In the third case, the lack of correlation between that factor and other factors leads to underidentification. (A one factor model with two indicators is not identified!).

It should be noted that "large loading" and "statistically significant loading" are not necessarily the same in this context. A factor which causes anomalies in the test sequence may have three or four statistically significant, but numerically low loadings!

Example

The application of the proposed test series for the analysis of MTMM matrices will now be demonstrated with data from a pilot study for the international research project Quality of Attitude Questions conducted in Austria, 1987. A sample of 179 persons from Vienna were asked about their "job satisfaction" (t_1), "satisfaction with their financial situation" (t_2), "satisfaction with their sexual situation" (t_3), "satisfaction with their social contacts" (t_4), and their "overall happiness" (t_5). The responses of each par-

table 3: Correlations among 15 answers of 179 respondents to questions concerning t=5 traits, each measured with m=3 methods

		x_1	x_2	x_3	x_4	x_5	x_6	x_7	x_8	x_9	x_{10}	x_{11}	x_{12}	x_{13}	x_{14}	x_{15}
m=1																
t_1	x_1	1														
t_2	x_2	.463	1													
t_3	x_3	.239	.274	1												
t_4	x_4	.383	.285	.347	1											
t_5	x_5	.445	.447	.536	.436	1										
m=2																
t_1	x_6	.826	.426	.233	.307	.402	1									
t_2	x_7	.456	.720	.141	.233	.372	.548	1								
t_3	x_8	.168	.125	.786	.263	.439	.275	.180	1							
t_4	x_9	.231	.175	.253	.556	.326	.314	.247	.370	1						
t_5	x_{10}	.396	.375	.513	.371	.665	.382	.399	.483	.359	1					
m=3																
t_1	x_{11}	.751	.388	.178	.359	.405	.821	.458	.193	.379	.430	1				
t_2	x_{12}	.420	.668	.205	.301	.422	.513	.721	.142	.234	.385	.553	1			
t_3	x_{13}	.257	.226	.726	.267	.490	.332	.211	.743	.341	.482	.362	.362	1		
t_4	x_{14}	.325	.260	.277	.650	.365	.385	.261	.340	.629	.384	.529	.453	.466	1	
t_5	x_{15}	.302	.305	.261	.202	.486	.312	.318	.271	.425	.590	.424	.448	.440	.390	1

ticipant to three (identical) questions were measured on each of these five dimensions. The three questions differed only with respect to the scales used: a five-point scale (m_1), a nine-point scale (m_2), and a magnitude scale (m_3).

The intercorrelations of the 15 indicators resulting from this research design are presented in table 3. All analyses were performed using the LISREL VI program (Jöreskog and Sörbom, 1983).

The results of the nested model comparisons are given in table 4. The outcomes presented in this table can be interpreted as follows:

1. Both trait and method factors uniquely explain a highly significant proportion of variance.

table 4: Nested model comparisons for the matrix in table 3

MODELS

M_0	specifications	χ^2	df	P	anomalies
H_0	no trait factors no method factors	1997.73	105	.000	-
H_1	t=5 traits only (oblique)	312.63	80	.000	-
H_2	t=5 (oblique) traits, m=3 (orthogonal) methods	107.12	65	.001	-
H_3	t=5 (oblique) traits, m (oblique) methods	157.44	65	.000	1
A	m=3 methods only (oblique)	1402.01	90	.000	-
B	1 general trait, m=3 (oblique) methods	915.95	75	.000	TD(10) < 0 [2]

MODEL COMPARISONS

comparison		testing for:	Q	accepted model
H_0	H_1	significance of trait factors	67.4	H_1
H_1	H_2	significance of method factors	13.7	H_2
H_2	H_3	discriminability of method effects	-	H_2
A	H_3	convergent validity	51.8	H_2
B	H_3	discriminant validity	80.8	H_2

[1] For model H_3 (t oblique traits, m oblique methods), iterations do not converge. Therefore, the method factor intercorrelations were fixed to 1 in order to test for the discriminability of method effects.

[2] Parameter not significant

2. The best fitting model H_2 with 5 oblique traits and 3 orthogonal methods (though statistically rejectable) is acceptable on practical grounds: Delta = .946.
3. The parameter estimates of H_2 cannot be fully trusted because the method factor intercorrelations are empirically underidentified. The reason for this problem lies in (mostly) significant but numerically low loadings in the nine-point scale method factor (m_2). The corresponding five loadings and their standard errors were estimated to be .156 (.063), .212 (.079), .274 (.086), .202 (.096), and -.097 (.111). So, the dilemma is that this factor cannot be deleted without substantially increasing the goodness-of-fit statistic, whereas including this factor leads to biased estimates of valid, method, and residual variance components, if the method factors are correlated.

CONCLUSIONS

There is no obvious simple solution for this dilemma, except the inclusion of additional real factors, which can make the iterative estimation procedure more stable. In the original analysis of the given MTMM-matrix (see Költringer and Kluscarits, 1988), five factors were added: four (perfectly measured) socio-demographic variables and one response set dimension (extreme score tendency). Within the resulting (good fitting) model, slightly higher method effects and rather high method factor intercorrelations were estimated: $r(m_1, m_2) = .549$ (.156), $r(m_1, m_3) = .498$ (.131), and $r(m_2, m_3) = .651$ (.105).

In general, the more traits and methods that are included in the MTMM-design, the lower the probability of less than three "large" loadings per factor, i.e., the greater the chance of detecting low or moderate method (or trait) variance components when actually only low or moderate method (or trait) effects are present.

REFERENCES

Alwin, D. (1974). An analytic comparison of four approaches to the interpretation of relationships in the multitrait-multimethod matrix. In H.L. Costner (Ed.), *Sociological methodology 1974.* San Francisco: Jossey-Bass.

Bentler, P.M., & Bonett, D.G. (1980). Significance tests and goodness of fit in the analysis of covariance structures. *Psychological Bulletin, 88,* 87-104.

Jöreskog, K.G. (1969). A general approach to confirmatory maximum likelihood factor analysis. *Psychometrica, 34,* 183-202.

Jöreskog, K.G. (1971). Statistical analysis of sets of congeneric tests. *Psychometrica, 36,* 109-133.

Jöreskog, K.G. (1974). Analyzing psychological data by structural analysis of covariance matrices. In C. Atkinson et al. (Eds.), *Contemporary developments in mathematical psychology.* San Francisco: Freeman.

Jöreskog, K.G., & Sörbom, D. (1983). *LISREL VI users guide.* University of Uppsala: Department of Statistics.

Kenny, D.A. (1976). An empirical application of confirmatory factor analysis to the multitrait-multimethod matrix. *Journal of Experimental Social Psychology, 12,* 247-252.

Költringer, R., & Kluscarits, W. (1988). *Subjektive Sozialindikatoren und Messfehlerproblematik: Ein kritischer Beitrag zur gegenwärtigen Praxis der Lebensqualitätsforschung.* Wien (unpublished).

Loehlin, J.C. (1987). *Latent variable models.* Hilldale, London: LEA.

Rindskopf, D. (1983). Parameterizing inequality constraints on unique variances in linear structural models. *Psychometrica, 48,* 73-83.

Rindskopf, D. (1984). Structural equation models: empirical identification, Heywood cases, and related problems. *Sociological Methods and Research, 13,* 109-119.

Schmitt, N., & Stults, D.M. (1986). Methodology review: Analysis of multitrait-multimethod matrices. *Applied Psychological Measurement, 10,* 1-22.

Widaman, K.F. (1985). Hierarchically nested covariance structure models for multitrait-multimethod data. *Applied Psychological Measurement, 9,* 1-26.

chapter 4

Robustness issues in the analysis of MTMM and RMM models

A. SATORRA

INTRODUCTION

Structural equation models are widely used in the social, behavioural and economic sciences to analyze linear relationships among variables. Overviews of the topic, with varied emphasis, can be found in the books of Aigner and Goldberger (1977), Goldberger and Duncan (1973), Saris and Stronkhorst (1984), or review articles such as those of Saris (1980), Anderson (1984), Jöreskog (1985) and Bentler (1986).

Estimation and evaluation of the fit of the model are the two major statistical tasks in structural equation modeling. This requires obtaining parameter estimates, i.e. fitting the model, and evaluating the sampling variability of the estimates as well as the null distribution of the test statistics. Assumptions are needed to justify the statistical conclusions of the analysis. These assumptions can be roughly divided into two types: structural and distributional. Structural assumptions determine a model of interrelationships among observed and/or theoretical variables, and imply a specific structure for the covariance matrix of the vector of observed variables. Distributional assumptions refer to the form of the distribution of the random vector of observable variables. Typically, the data analyzed is supposed to be composed of independent and identically distributed (iid) realizations of a p-dimensional random vector **z**, whose distribution is often assumed to be multivariate normal. Distributional assumptions play a crucial role in the assessment of the precision of the estimators, and in the calculation of the significance level ("p-values") of test statistics. In

This research was made possible by a grant from the Dutch Organization for Advancement of Pure Research (ZWO). The author wishes also to express his gratitude to Willem E. Saris whose comments and suggestions have enhanced enormously the presentation of this paper.

practice, data tend to be skewed and non-normal, thus the normality assumption for **z** has been subject to criticism.

The restrictive character of the normality assumption has led to the development of asymptotic distribution-free methods. These methods do not assume a specific distribution for **z** and produce results which are valid (asymptotically) under a wide variety of distributions of the data; moreover, they are asymptotically optimal (even if it turns out that the data is in fact normal). However, these distribution-free methods have serious disadvantages when used in practice: they are computationally expensive and, more importantly, they lack robustness with moderate to small sample sizes. Thus, the normality assumption still plays a major role in the practice of structural equation modeling.

Violation of assumptions may of course invalidate some conclusions of the analysis. Effectively, researchers draw conclusions on the assumption that estimators are consistent, asymptotically normal and with a variance that can be approximated from the data; also, the researcher hopes that the (null) distribution of a test statistic will adjust to a specific theoretical distribution, for instance the (centred) χ^2. It has been recognized that violation of structural or distributional assumptions may give rise to different types of distortion (e.g., Satorra, 1989). For instance, when the (structural) assumption that the model holds is violated, consistency of the parameter estimators may be lost; while violation of a distributional assumption, such as the normality assumption for **z**, does not distort the consistency of the parameter estimates, nor their characteristic of being asymptotically normally distributed. The latter may, of course, invalidate the normal theory variances of such estimates.

The elucidation of structural misspecification is one of the goals of the typical goodness-of-fit test statistics. However, the significance of a goodness of fit statistic should be due to model misspecification, and not just improper specification of the distribution of **z**. Thus, in the evaluation of a model, the distortion of the asymptotic distributions of test statistics by distributional misspecification is a matter of concern.

In this chapter we concentrate on issues related to distributional misspecification. To simplify the discussion, it will be assumed throughout that the model is not misspecified in the structural sense; i.e., the linear equations that constitute the model specification correspond unequivocally with the data generating mechanism. It should be noted, however, that most of the theory discussed below extends easily to the case of "slight" structural misspecification of the model (Satorra, 1989). Special attention will be paid to assessing the consequences of distributional misspecification.

The theoretical discussion will be restricted to two types of models used in comparative survey research, where the effects of different ways of formulating questions is the subject of investigation. The two models considered are the classical multitrait multimethod (MTMM) model introduced by Campbell and Fiske (1959), and the repeated multitrait multimethod (RMM) model discussed by Saris in this volume. The general consequences of analyzing these type of models under some form of distributional misspecification will be discussed.

This chapter contains the following:

- a presentation of the general class of linear latent variable models
- a review of the covariance structure analysis within a (weighted) least squares approach (which encompasses most of the estimation methods used in practice)
- the development of the asymptotic distribution of estimators and test statistics
- consideration of an asymptotic distribution-free estimate for the matrix Γ of asymptotic covariances of sample covariances (which will be shown to play a crucial role in assessing sampling variability)
- a description of the covariance structure analysis under the assumption that $\mathbf{z}$ is normally distributed
- the presentation of a theory for asymptotic robustness of the normal theory results
- an application of this theory to MTMM and RMM types of models

MODELS: LINEAR RELATIONS

The models we consider state that a vector $\mathbf{z}$ of observable variables is structured as a linear function of more basic variables, some of which may be latent (unobservable).

A typical example is the factor analysis model

$$\mathbf{z} = \Lambda \xi + \varepsilon \tag{1}$$

where Λ is a matrix of parameters and ξ and ε are respectively common and specific (unobservable) factors, which are assumed to be uncorrelated. Without loss of generality, we can also assume that ξ and ε are centred variables. Note that the "weak" assumption of uncorrelation, and not the "stronger" assumption of independence, is typically considered in specifying the factor model. (Independence, and not uncorrelation, will be the critical condition for the robustness results to be described in the next section.)

Another typical model is the structural equation model, or "path analysis" model, used by econometricians and other social scientists. In this model the vector $\mathbf{z}$ of observable variables can be divided into endogenous and exogenous variables, vector $\mathbf{y}$ and $\mathbf{x}$ respectively, in such a way that the following linear equation holds:

$$\mathbf{y} = \mathbf{B}\mathbf{y} + \Gamma\mathbf{x} + \mathbf{u} \tag{2}$$

where $\mathbf{u}$ is a vector of (unobservable) equation "disturbances", and $\mathbf{B}$ and Γ are parameter matrices. Uncorrelation (again, non-necessarily independence) between $\mathbf{x}$ and $\mathbf{u}$ needs to be assumed.

A very general multivariate relation is the LISREL model (e.g., Jöreskog, 1977). LISREL combines equations of types (1) and (2), i.e. measurement and structural equations, as follows

$$\mathbf{y} = \Lambda_y\eta + \varepsilon \tag{3a}$$

$$\mathbf{x} = \Lambda_x\xi + \delta \tag{3b}$$

$$\eta = \mathbf{B}_o\eta + \Gamma\xi + \zeta \tag{3c}$$

where $\mathbf{y}$ and $\mathbf{x}$ are observable variables; ε, δ, ξ and ζ are uncorrelated random vectors (usually unobservable) with finite covariance matrices Θ_ε, Θ_δ, Φ and Ψ, respectively; and Λ_y, Λ_x, $\mathbf{B}_o$ and Γ are matrices containing the unknown parameters of the model. Without loss of generality, it can be assumed that δ, ξ and ζ have a mean equal to zero (hence η, $\mathbf{y}$ and $\mathbf{x}$ are also centred variables) and that $(\mathbf{I} - \mathbf{B}_o)$ is a non-singular matrix.

Information with regard to the form of the matrices $\mathbf{B}_o$, Λ, Φ and Ψ will be added to (3) in order to obtain specific models. In LISREL the elements of the parameter matrices can be restricted to values specified as equal, or can be restricted by equalities among them. The distinct and functionally independent unknown elements of $\mathbf{B}_o$, Λ_y, Λ_x, Φ and Ψ will be assembled into a q-dimensional parameter vector θ.

For the approach used in this chapter, we can rewrite the models in Bentler's (1983) general formulation

$$\mathbf{z} = \mathbf{A}\xi, \tag{4}$$

where **A** is a parameter matrix which is a function of some basic parameters, which are components of θ, and ξ is a vector composed of observable and latent variables. The covariance matrix of ξ will be denoted by Φ and is also assumed to be a function of θ.

The matrix A can be partitioned as $A \equiv [\, A_1 \mid A_2 \mid \dots \mid A_k \,]$, conforming with the partition $\xi \equiv [\, \xi_1' \mid \xi_2' \mid \dots \mid \xi_k' \,]'$ of ξ, in such a way that the ξ_k's are uncorrelated with each other; the covariance matrix of each ξ_k, k = 1, 2, ..., K, will be denoted by Φ_k. Thus (4) can be written as

$$\mathbf{z} = [\, \mathbf{A}_1 \mid \mathbf{A}_2 \mid \dots \mid \mathbf{A}_k] [\, \xi_1' \mid \xi_2' \mid \dots \mid \xi_k' \,]' \tag{5}$$

with the matrix Φ of variances and covariances of $[\, \xi_1' \mid \xi_2' \mid \dots \mid \xi_k' \,]$ having the block diagonal form

$$\Phi = \text{diag}\, [\, \Phi_1 \mid \Phi_2 \mid \dots \mid \Phi_k \,] \tag{6}$$

Typically, the Φ_k's will be unrestricted symmetric matrices (i.e., the model will specify each Φ_k to be a symmetric and "free" matrix), and the "blocks" Φ_k's will be expressed as a (smooth) function of the parameter vector θ.

For example, model (1) can be written as

$$\mathbf{z} = [\, \Lambda \mid \mathbf{I} \,] [\, \xi' \mid \varepsilon' \,]' \tag{7}$$

the path analysis model (2) as

$$\mathbf{z} \equiv \begin{bmatrix} \mathbf{y} \\ \mathbf{x} \end{bmatrix} = \begin{bmatrix} (\mathbf{I}-\mathbf{B})^{-1}\Gamma & (\mathbf{I}-\mathbf{B})^{-1} \\ \mathbf{I} & \mathbf{O} \end{bmatrix} \begin{bmatrix} \mathbf{x} \\ \mathbf{u} \end{bmatrix} \tag{8}$$

and the general LISREL model (3) as

$$\mathbf{z} \equiv \begin{bmatrix} \mathbf{y} \\ \mathbf{x} \end{bmatrix} = \begin{bmatrix} \Lambda_y(\mathbf{I}-\mathbf{B})^{-1}\Gamma & \Lambda_y(\mathbf{I}-\mathbf{B})^{-1} & \mathbf{I} & 0 \\ \Lambda_x & 0 & 0 & \mathbf{I} \end{bmatrix} \begin{bmatrix} \xi \\ \zeta \\ \varepsilon \\ \delta \end{bmatrix} \tag{9}$$

(Throughout this chapter, **I** and **0** denote respectively identity and zero matrices of appropriate dimension.)

Each of the above linear relations implies a specific structure for the covariance matrix Σ of **z**; that is, the structural model expressed by the linear equations gives rise to a covariance structure for Σ, i.e. to a p x p-dimensional matrix-valued function $\Sigma = \Sigma(\theta)$ of θ (this function is generally non-linear, but smooth enough to be twice continuously differentiated). The parameter vector θ varies in the parameter space Θ (a q-dimensional compact subset of $\mathbf{R}^q$). For example, (1) and the assumption that ξ and ε are uncorrelated, imply the following typical covariance structure for the factor model:

$$\Sigma = \Lambda \Phi \Lambda' + \Psi \tag{10}$$

where the matrices Λ, Φ and Ψ contain the parameters of the model.

WEIGHTED LEAST SQUARES ANALYSIS OF COVARIANCE STRUCTURES

Let $\Sigma = \Sigma(\theta)$, θ in Θ, be a covariance structure, and suppose that n observations, $\mathbf{z}_1, \mathbf{z}_2, \ldots, \mathbf{z}_n$, of **z** are available. Typically, the analysis starts with the computation of the matrix of sample covariances

$$\mathbf{S} = \frac{1}{n-1} \sum_{i=1}^{n} (\mathbf{z}_i - \bar{\mathbf{z}})(\mathbf{z}_i - \bar{\mathbf{z}})' \tag{11}$$

where $\bar{\mathbf{z}} \equiv \frac{1}{n} \sum_{i=1}^{n} \mathbf{z}_i$

It is well known that, under very mild conditions, namely the assumption that **z** has finite fourth-order moments and the sample is composed of n iid realizations of **z**, the matrix **S** converges (in probability) to the true (population) covariance matrix of **z**, say Σ_o.

The statement that the model holds will mean precisely that $\Sigma_o = \Sigma(\theta_o)$ for a certain (usually unknown) true vector of parameters θ_o (which, for technical reasons, will be assumed to be an interior point of the parameter space Θ).

Within this framework, it seems natural to estimate the parameters of the model by choosing the value of θ, say $\hat{\theta}$ for which the fitted covariance matrix $\Sigma \equiv \Sigma(\hat{\theta})$ is "closest" to the observed sample covariance matrix **S**. Of course, when a perfect fit of $\Sigma(\hat{\theta}) = \mathbf{S}$ is not attainable (that is, when the model is overidentified; i.e., there are more distinct elements in q than in S), different scales for measuring "closeness" will yield alternative parameter estimates. The quality of these alternative estimates can vary.

Following a standard approach in statistics, the fitting of $\Sigma(\theta)$ to **S** can be framed into a least squares approach. Let us arrange the distinct elements of **S** and Σ in a column vector; that is, let us use $\mathbf{s} = \text{vech}(\mathbf{S})$ and $\sigma(\theta) = \text{vech}\ (\Sigma(\theta))$, where vech (..) lists the non-duplicated elements of a symmetric matrix which are on and below the diagonal (starting with the first column), and consider the following weighted least squares (WLS) fitting function:

$$\mathbf{F}(\theta) = (\mathbf{s} - \sigma(\theta))'\mathbf{V}(\mathbf{s} - \sigma(\theta)) \qquad (12)$$

with **V** a specific positive definite weight matrix. The criterion of $\hat{\sigma} \equiv \text{vech}(\Sigma(\hat{\theta}))$ being close to **s** leads naturally to choosing $\hat{\theta}$ to minimize (12). Different matrices **V** will induce different ways of measuring "closeness" and may lead to different estimates. The WLS approach is typical in covariance structure analysis (e.g., Browne, 1974 and 1984; Bentler and Dijkstra, 1985; Fuller, 1987, section 4.2).

We assume throughout that enough regularity conditions hold for the (WLS estimator) $\hat{\theta}$ which minimizes (12) to be uniquely defined (that is, we assume that the model is identified). The matrix **V** will often be a function of n, i.e. $\mathbf{V} \equiv \mathbf{V}_n$. In that case, we need to assume that $\lim_{n\to\infty}\mathbf{V}_n = \mathbf{V}$, a positive definite matrix (when $\mathbf{V}_n$ is stochastic, then "lim" will be understood to denote limit in probability).

WLS fitting functions, with a variety of weight matrices **V**, are implemented in typical computer programs for covariance structure analysis, such as LISREL (Jöreskog and Sörbom), EQS (Bentler, 1985), LISCOMP (Muthén, 1987), etc. A specific case of WLS is the "unweighted" least squares (ULS) analysis for which $\mathbf{V} = \mathbf{I}$. (The latest version of LISREL also implements a diagonal weighted least squares (DWLS) analysis). The (asymptotic) properties of the estimators and the test statistics may vary for different combinations of the weight matrix **V** and asymptotic distribution of vector **s** of sample covariances. A typical choice is to use a weight matrix **V** such that the analysis is optimal when the distribution of **z** is normal (or, equivalently, when the sample covariance matrix **S** follows a Wishart distribution). In fact, it has been shown (Browne, 1974) that normal theory maximum likelihood minimizes the following fitting function :

$$F_{ML}(\theta) = \log |\Sigma(\theta)| + \text{trace}\{S\Sigma(\theta)^{-1}\} - \log |S| - p \qquad (13)$$

and gives (with a probability of one) the same estimates as those provided by a WLS fitting function with weight matrix **V** which is a specific function of Σ (e.g., Browne, 1974; Lee and Jennrich, 1979; Fuller, 1987). Later in this chapter we will see that there are a variety of WLS estimators with the same asymptotic properties as the ML estimator that minimizes (13).

Shapiro (1985) has shown that most of the estimators in covariance structure analysis are asymptotically equivalent (up to first-order asymptotic expansions) to WLS estimators (see also, Satorra, 1989). Thus the extent to which discussion in this chapter is confined to WLS analysis should not be seen as restrictive.

An important aspect of covariance structure analysis is the evaluation of the fit of the model. A typical statistic for such evaluation is $n(\hat{F}) \equiv nF(\hat{\theta})$, which is the so called χ^2 goodness of fit statistic. The residual vector $\mathbf{d} \equiv \text{vech}(\mathbf{S} - \Sigma(\hat{\theta}))$ also provides information for the fit of the model. (Alternative test statistics used in structural equation modeling are reviewed in Satorra, 1989.)

CONSISTENCY AND ASYMPTOTIC NORMALITY OF WLS ESTIMATORS

Note that (12) and (13) attain the minimum (of zero) when **S** equals Σ. Also, when the model holds, **S** converges in probability to $\Sigma^o = \Sigma(\theta^o)$. This implies that, as n increases, $\hat{\theta}$ converges in probability to θ^o, $\hat{\theta}$ is a consistent estimator of the true parameter value θ^o. Consistency holds regardless of the type of "weight" matrix **V** used in (12) and regardless of the type of distribution of **z**. For instance, a ULS analysis, i.e. when **V** is the identity matrix, produces estimators which may not be the best in terms of efficiency (i.e., in having the minimum variance among alternative estimators), but they are consistent. Thus, consistency holds given any matrix **V** and distribution of **z** (it is a "robust" property of the estimators) whenever the model holds. Of course, consistency may be lost under structural misspecification.

Further, it is usually desirable to assess the sampling variability of the estimates and test statistics. That is, we want to know the precision of $\hat{\theta}$ as an estimator of the true value θ^o; moreover, we also want to compute the significance level ("p-value") of a test statistic such as $n\hat{F}$. To attain this information we must consider the specific distribution of **z**.

Under very weak conditions, namely the assumptions of a sample of iid observations of **z**, and the existence of fourth-order moments of **z**, i.e. within a "distribution-free" context, the following asymptotic result holds (see, e.g., Muirhead, 1982):

$$\sqrt{\mathbf{n}}\,(\mathbf{s} - \sigma^o) \rightarrow N(0,\Gamma) \tag{14}$$

where "$\rightarrow$" denotes convergence in distribution, $N(0, \Gamma)$ is a multivariate normal distribution of zero mean and covariance matrix Γ, and $n^{-1}\Gamma \equiv \text{avar}(\mathbf{s})$ is the matrix of asymptotic covariances of **s**. (To simplify the discussion we will not deal with the case of "slight" structural misspecification in which the normal distribution on the right hand side of (14) has a non-zero mean; this case is discussed in Satorra, 1989). The result (14) of asymptotic normality of **s**, the fact that θ can be viewed locally as a function $\hat{\theta} = \theta\,(\mathbf{s})$ of **s** (not necessarily expressible in closed form, but satisfying typical regularity conditions of differentiability, as follows from $\hat{\theta}$ being a minimizer of (11) and (12)), and, finally, the so-called "delta-method" (Rao, 1965), leads immediately to $\hat{\theta}$ being asymptotically normal, with an (asymptotic) covariance matrix (Ferguson, 1958).

$$\text{avar}\ (\hat{\theta}) = \frac{1}{n}\,(\Delta'\mathbf{V}\Delta)^{-1}\,\Delta'\mathbf{V}\Gamma\mathbf{V}\Delta\,(\Delta'\mathbf{V}\,\Delta)^{-1} \tag{15}$$

where the derivative matrix $\Delta \equiv \partial\,\sigma/\,\partial\theta'$ is evaluated at θ^o (or at $\hat{\theta}$ when a consistent estimate is required).

Further, with respect to the χ^2 goodness of fit statistic, it holds that (Satorra and Bentler, 1986)

$$\mathbf{nF}(\hat{\theta}) \rightarrow \sum_{i=1}^{r} \alpha_i \mathbf{T}_i \tag{16a}$$

in distribution, where the $\mathbf{T}_i$'s are independent χ^2 distributions with 1 degree of freedom (df) and the α_i's are the r non-zero eigenvalues of the following product matrix

$$(\mathbf{V} - \mathbf{V}\Delta\,(\Delta'\mathbf{V}\,\Delta)^{-1}\,\Delta'\mathbf{V})\Gamma \qquad (16b)$$

That is, for any type of weight matrix used, and for any type of distribution of the data (as long as the fourth-order moment of **z** are finite), the vector of estimators $\hat{\theta}$ is asymptotically normally distributed; and, when the model holds, the goodness of fit statistic $\mathbf{nF}(\hat{\theta})$ is asymptotically distributed as a mixture of (centered) χ^2 distributions each with 1 df.

To approximate (15) and (16) in a specific application, consistent estimates of population matrices will be required. Note that a consistent estimate for Γ (i.e. n^{-1} times the matrix of asymptotic variances and covariances of **s**) is required if we want to evaluate the asymptotic variances of $\hat{\theta}$, or the null distribution of $\mathbf{nF}(\hat{\theta})$. An appropriate (non-parametric) consistent estimate for Γ, based on the fourth-order sample moments, can easily be constructed (see the section below). Satorra and Bentler (1988a) have shown how to implement (16) in order to obtain an appropriate significance level for goodness-of-fit statistics.

It should be noted that in the specific case where V (or the probability limit of Vn) is equal to Γ^{-1}, then instead of (15) and (16) we have the simpler expression

$$\text{avar}\ (\hat{\theta}) = \frac{1}{n}\,(\Delta'\mathbf{V}\,\Delta)^{-1} \qquad (17)$$

for the matrix of asymptotic variances of $\hat{\theta}$, and an asymptotically exact χ^2 distribution,

$$\mathbf{nF}(\hat{\theta}) \rightarrow \chi^2{}_r \qquad (18)$$

for $\mathbf{nF}(\hat{\theta})$. Here $r \equiv p(p + 1)/2 - q$ are the corresponding degrees of freedom.

Furthermore, in the case of $\mathbf{V} = \Gamma^{-1}$ it can be shown that the matrix avar $(\hat{\theta})$ attains its minimum within the class of estimators that minimize a WLS function such as (12); that is, efficiency is attained within the class of WLS estimators (e.g., Shapiro, 1986). When this occurs, the weight matrix **V**, or the corresponding discrepancy function, is said to be asymptotically optimal (AO)

(Satorra, 1989). Note that AO is a property that requires a specific relation between the weight matrix **V** and the asymptotic distribution of **s**.

Note that in case of an AO weight matrix **V**, the expression (17) for the matrix of asymptotic covariances of θ can be written as

$$\text{avar}\,(\hat{\theta}) = \frac{1}{n}\,(\Delta'\Gamma\,\Delta)^{-1} \qquad (19)$$

which shows that it depends on the model (through Δ) and the asymptotic distribution of **s** (through Γ). As (19) gives the variances for the more precise estimates (within the class of WLS estimates), the asymptotic variances of the estimates extracted from this matrix will be denoted AO variances (note that they are uniquely defined given the model and the asymptotic distribution of **s**). Again, note that this optimality in estimation is attained when the chosen weight matrix **V** is the inverse of Γ, which in turn depends on the distribution of the observable variables **z**.

In theory, asymptotic optimality can be easily achieved for a general class of distributions of **z** by using the so-called asymptotic distribution-free methods, which require only the finiteness of the eighth-order moments of **z**. However, the use of such methods leads to serious practical problems such as high computational requirements and also lack of robustness with small sample sizes.

ASYMPTOTIC DISTRIBUTION-FREE ESTIMATE OF Γ

Under very general conditions (namely, existence of eighth-order moments of **z**), a consistent estimator for matrix Γ of (14) is the following matrix of fourth-order sample moments (Fuller, 1987, pp. 332):

$$\mathbf{H} = \frac{1}{n}\sum_{i=1}^{n}(\mathbf{b}_i - \bar{\mathbf{b}})(\mathbf{b}_i - \bar{\mathbf{b}})' \qquad (20)$$

where

$$\mathbf{b}_i = \text{vech}\,(\mathbf{z}_i - \bar{\mathbf{z}})(\mathbf{z}_i - \bar{\mathbf{z}})' \qquad (21)$$

and $\bar{\mathbf{z}}$ and $\bar{\mathbf{b}}$ are the vectors of sample means of the $\mathbf{z}_i$'s and $\mathbf{b}_i$'s respectively.

Note that a WLS analysis with $\mathbf{V}_n \equiv \mathbf{H}^{-1}$ will satisfy the condition of $\mathbf{V}_n$ converging to Γ^{-1}; hence it will yield an AO analysis regardless the type of distribution of $\mathbf{z}$. This is the so-called asymptotic distribution-free (ADF) analysis of Browne (1982, 1984) and Bentler (1983); since it is a weighted least squares analysis with the weight matrix based on the inverse of an asymptotic distribution-free estimate of Γ, it will be denoted ADF-WLS.

However, the ADF-WLS has some practical drawbacks. The number of distinct (fourth-order sample moments) elements of Γ increases at the fourth power of the dimension p of $\mathbf{z}$ and thus even for moderate values of the dimension p of $\mathbf{z}$, a matrix of very high dimension must be stored and inverted; thus, ADF-WLS may become computationaly unfeasible for moderate-to-large size models. Furthermore, stable estimates of higher-order moments require large sample sizes, hence the ADF-WLS analysis may lack robustness with small sample size. The major property of ADF-WLS, i.e. being AO under a wide range of distributions of the data, may then just be an asymptotic quality requiring a larger sample size than can be obtained.

As an alternative to the AO analysis implied by ADF-WLS, we could use any type of WLS as long as we take care to implement correct asymptotic variances for $\hat{\theta}$ and asymptotically correct null distributions for the test statistics. This would imply using (15) and (16) with $\mathbf{H}$ substituted for Γ and, as usual, evaluating the matrix of derivatives Δ at $\hat{\theta}$ (Satorra and Bentler, 1988a and 1988b). Note that this approach would not require inverting the matrix $\mathbf{H}$. Also, such an approach produces (asymptotic) variances for the ML estimators, which are correct even for non-normal data. Of course, theoretical arguments may be used to claim that such "ML" analysis with non-normal data is asymptotically less efficient than an ADF analysis; however, as has been mentioned above, for practical moderate-to-small sample sizes such asymptotic characteristic may not be achieved.

A drastic reduction of the computational complexity and sample instability is obtained when the higher-order moments (here the fourth-order moments of Γ) are expressed as a function of the considerably fewer (and certainly more stable with regard to sampling fluctuations) second-order moments of matrix Σ; this occurs when the distribution of $\mathbf{z}$ is multivariate normal.

INFERENCES BASED ON THE NORMALITY ASSUMPTION

A more restrictive assumption than just finiteness of the eighth-order moments is the assumption that **z** is normally distributed. Under this assumption, the elements $\Gamma_{ij,\,kl} \equiv \mathbf{n}\ \mathrm{acov}(\mathbf{s}_{ij}, \mathbf{s}_{kl})$ of matrix Γ of asymptotic covariances of **s** are

$$\Gamma_{ij,\,kl} = \sigma_{ik}\,\sigma_{jl} + \sigma_{il}\,\sigma_{jk} \tag{22}$$

that is, the elements of Γ are simply a function of the second-order moments of **z** (variances and covariances). Thus, when **z** is normally distributed, a consistent estimate for $\Gamma_{ij,\,kl}$ will be $(\sigma_{ik}\,\sigma_{jl} + \sigma_{il}\,\sigma_{jk})$. In matrix notation, (22) gives the following normal theory (NT) form of Γ:

$$\Gamma = 2\,\mathbf{D}^{+}(\Sigma \otimes \Sigma)\mathbf{D}^{+\prime} \tag{23}$$

with "$\otimes$" denoting the Kronecker product and $\mathbf{D}^{+}$ the Moore-Penrose inverse of the (fixed and known) "duplication" matrix **D** (for result (23), and details on the duplication matrix, see Magnus and Neudecker, 1986). Hence, under the assumption of normality, a matrix expression for a consistent estimate of Γ is given by the right hand side of (23) with **S**, or other consistent estimators of Σ, e.g.$\hat{\Sigma}$, substituting for Σ. Accordingly, the expressions for the NT matrix of asymptotic variances of $\hat{\theta}$ and asymptotic distribution of $\mathrm{n}F(\hat{\theta})$, will be obtained from (15) and (16), respectively, with Γ taking the NT form of (23).

Consider now a WLS analysis with weight matrix

$$\mathbf{V} \equiv \frac{1}{2}\,\mathbf{D}'(\Sigma^{-1} \otimes \Sigma^{-1})\mathbf{D} \tag{24}$$

(or $\mathbf{V}_n$ converging to the above form of **V**), the inverse of the matrix Γ given in (23). If, in addition, **z** is normally distributed, then this WLS analysis will be AO (since the condition $\mathbf{V} = \Gamma^{-1}$ is trivially satisfied). An obvious implementation of (24) would substitute Σ for a consistent estimate, such as **S** (or $\hat{\Sigma}$). Since the weight matrix (24) is the inverse of Γ under the normal theory assumption for **z**, a WLS analysis with **V** as in (24) will be denoted NT- WLS (note that NT-WLS is AO under normal theory of **z**).

Such NT-WLS is not unique as there are different sets of $\mathbf{V}_n$ converging to the **V** of (24); however, alternative NT-WLS give rise to asymptotically equivalent statistics, which are also equivalent to those obtained when minimizing the ML fitting function (13) (e.g., Satorra, 1989). Note also that, regardless of the distribution of **z**, the estimators will be consistent and asymptotically normally distributed. Note that the matrix of asymptotic variances of $\hat{\theta}$, and the asymptotic null distribution of $nF(\hat{\theta})$, which are correct even for non-normal **z**, derive from (15) and (16) respectively with **H** of (20) substituted for Γ.

A special situation occurs when NT-WLS is used and **z** is in fact normally distributed. As condition $\mathbf{V} = \Gamma^{-1}$ will then hold, the analysis will be AO. Hence, in such a case, where **z** is normally distributed, the asymptotic variance of $\hat{\theta}$ follows from the reduced expression (17), and the asymptotic null distribution of $nF(\hat{\theta})$ will be the exact χ^2 distribution of (18). Furthermore, it can be shown that for such a combination of normality of **z** with a NT-WLS fitting function, the estimates obtained are asymptotically equivalent to maximum likelihood estimates, hence asymptotic efficiency is attained (Browne, 1974).

The above case of using a normal-theory-based fitting function with normally distributed data will not often occur in practice. The circumstances of each individual case may lead to the use of a simpler fitting function than a NT-WLS (e.g., $\mathbf{V} \equiv \mathbf{I}$ may be used) or, more frequently, the actual distribution of **z** may deviate considerably from a multivariate normal distribution.

With regard to the parameters of most interest and $nF(\hat{\theta})$, the asymptotic distributions implied by (15) and (16) with Γ taking the normal theory form of (22), will be seen to be correct even for non-normal **z**, provided some model conditions hold. The general theory for these results can be found in Satorra and Bentler (1988b). An overview is given in the next section.

ASYMPTOTIC ROBUSTNESS OF NORMAL THEORY INFERENCES

In the discussion so far, the specification of the model at the level of linear equations, as discussed in section 2, was important only in deriving a specific covariance structure, $\Sigma = \Sigma(\theta)$, for the population matrix Σ of covariances of **z**. The distribution of **z** played a role only in deriving the matrix Γ, which characterizes the asymptotic distribution of vector **s** of sample covariances. As mentioned above, when **z** is normally distributed, Γ can be expressed as a function of Σ. In this section, the specification of the model at the level of the (structural) linear equations will be exploited further. It will be shown that in order to compute the asymptotic variances of the estimates, and (asymptotic) null distribution of the goodness-of-fit statistic (that is, in order to assess the variability of the statistics), we can simply use the NT form for Γ (that is, the expression of Γ given in (23)) even when **z** is non-normal and hence (23) is false. This section is a summary of earlier work by Satorra and Bentler (1988b), to which the reader is referred for full details.

To deal with this issue, it will be helpful to consider the concept of a distribution or quantity which is distribution-free of a specific random constituent of the model.

Definition (cf. Satorra and Bentler, 1988b)

The asymptotic distribution of a statistic, or a specific quantity, will be said to be distribution-free (DF) of a certain variable ξ, which has covariance matrix Φ, if that distribution or quantity depends on ξ through Φ (i.e., as when ξ is normally distributed).

As a specific example, consider **x** and **y** to be two independent random variables with finite second-order moments, and let s_{xy} be the (statistic) sample covariance based on n iid realizations of the 2-dimensional vector $\mathbf{z} \equiv (\mathbf{y},\mathbf{x})$. It is well known that the asymptotic distribution of $\sqrt{n}\, \mathbf{s}_{xy}$ is normal with a variance equal to the product $\sigma_{xx}\sigma_{yy}$ of variances of **x** and **y** respectively. Since this result holds regardless of the distribution of **x** and **y**, we will say that $\mathbf{s}_{xy}$ is asymptotically DF of both variables **x** and **y**.

Theorem 1 (cf. Corollary 3.2 of Satorra and Bentler, 1988b)

$$\text{Let } \mathbf{z} = [\, \mathbf{A}_1 \mid \mathbf{A}_2 \mid \dots \mid \mathbf{A}_k] \, [\, \xi_1' \mid \xi_2' \mid \dots \mid \xi_k' \,]'$$

where the $\mathbf{A}_k$ matrices are (twice continuously differentiable) functions of parameters of the model, and the random vectors ξ_k's are mutually independent with finite covariance matrices, Φ_{kk}'s. Assume further that each ξ_k which is non-normally distributed has an unrestricted covariance matrix Φ_{kk} which is unrelated to the $\mathbf{A}_k$'s and other covariance matrices Φ_{kk}'s. Then the χ^2 goodness of fit statistic $n\hat{F}$, the residual matrix $(\mathbf{S}-\hat{\Sigma})$, as well as the estimators of those parameters which are not elements of Φ_{kk}'s of non-normal ξ_k's, are asymptotically DF of **z**. Moreover, the AO variances for estimates of parameters which are not covariances of non-normal ξ_k's are also DF of **z**.

The proof of this theorem is given in Satorra and Bentler (1988b). Note that the covariance matrices Φ_{kk}'s of normal ξ_k's can also be functions of parameters of the model. However, for the conditions of the theorem to hold, such matrices can not be related to variances and covariances of non-normal ξ_k's.

The proof of the theorem hinges on the fact that linearity of the equations of the model allows us to express the statistics of interest (i.e., $\hat{\mathbf{F}}$, the residual matrix $(\mathbf{S} - \hat{\Sigma})$, and estimates of parameters that are not variances and covariances of non-normal ξ_r's) as being asymptotically equal to linear or quadratic functions of sample variances and covariances among the ξ_r's. It is crucial that the conditions of the theorem guarantee that non-normal ξ_r's contribute to the stochastic behaviour of the statistics of interest only through terms of type $\mathbf{S}_{\xi_r \xi_s}$ (sample covariances) with $r \neq s$; these are covariances among independent variables and thus are asymptotically DF of both variables ξ_r and ξ_s.

As implications of the theorem, note that when the asymptotic distribution of a statistic is DF of **z**, then we can use the formulae corresponding to the case of **z** being normally distributed. Thus, in order to compute the asymptotic variances of estimators and p-values for test statistics, in the corresponding expressions (i.e., (19) and (20)) we can use the NT form for Γ as given by (23). This will give an appropriate assessment of the variability of the statistics given any specific WLS fitting function, and not only those in the NT-WLS class. When the NT form for Γ gives correct asymptotic distributions even for non-normal **z**, then asymptotic robustness holds.

Note that our results on asymptotic robustness derive from the model specification and are not only specific to a given class of WLS functions. More specifically, the results of Satorra and Bentler mentioned above have implications not only for computing standard errors when using ML or NT-WLS fitting functions, but also when computing standard errors and p-values in the case, for instance, of an unweighted least squares (ULS) analysis when **z** is not necessarily normal.

Within a NT-WLS analysis, when the conditions of theorem 1 are satisfied, from the results just described it follows that the usual asymptotic variances of most of the estimates of interest, and χ-squaredness of $n\hat{\mathbf{F}}$, will also be correct even for non-normal data. The asymptotic robustness of the ML or NT-WLS analysis is due to Anderson and Amemiya (1988) and Anderson (1987, 1989); very similar results to those of Anderson can also be found in

Shapiro and Browne (1988), and Mooijaart and Bentler (1987). Note that theorem 1 (that is, theorem 3.3 of Satorra and Bentler, 1988b) extends the work mentioned to a broader class of fitting criteria than just the ML or NT-WLS type of analyses. Moreover, it also extends the robustness theory to inferences based on general types of statistics, such as the residual matrix ($\mathbf{S} - \hat{\Sigma}$) (to appreciate this fully, the reader should refer to the work of Satorra and Bentler mentioned above).

The above theorem ensures also that the AO variances of the estimates of some parameters will be DF of $\mathbf{z}$; i.e., such estimates of variance will be the same as when $\mathbf{z}$ is normally distributed, thus they can be obtained from

$$\text{avar}(\hat{\theta}) = \frac{1}{n}(\Delta'\mathbf{V}\Delta)^{-1} \tag{25}$$

with $\mathbf{V}$ taking the normal theory form (24). The variances extracted from (25) are, however, the conventional variances given by the standard programs when working with the specification of an ML or NT-WLS analysis. Hence, asymptotic robustness implies also that ML (or NT-WLS) gives optimal estimates for parameters other than the variances and covariances of non-normal ξ_r's (cf., Browne and Shapiro (1988)).

ROBUSTNESS RESULTS FOR MTMM AND RMM MODELS

The MTMM model

We consider a model for multitrait multimethod data (Campbell and Fiske, 1959). Each of a set of J concepts (traits or factors) is measured with a set of I different response scales (methods). The multitrait multimethod (MTMM) model specifies that the response $R(j,i)_\alpha$ on individual α for combination of factor F(j) and method M(i) satisfies the linear relation

$$R(j,i)_\alpha = \beta(j,i)\, F(j)_\alpha + \gamma(j,i)U(i)_\alpha + E(j,i)_\alpha, \quad \alpha = 1, 2, \ldots\ldots, N$$

where $F(j)_\alpha$, $U(i)_\alpha$ and $E(j,i)_\alpha$ are respectively the values on individual α of factor, method and error effects, and $\beta(j,i)$ and $\gamma(j,i)$ are parameter values.

By considering the vector-variables $R \equiv (R(1,1), R(2,1), \ldots, R(J,I))$, $F \equiv (F(1), F(2), \ldots, F(J))$, $U \equiv (U(1), U(2), \ldots, U(I))$ and $E \equiv$

(E(1,1), E(2,1), ..., E(J,I)), corresponding to responses, factors, methods and residuals, respectively, we can write the MTMM model as:

$$\mathbf{R} = \left[\begin{array}{c:c:c} \beta\text{'s} & \gamma\text{'s} & \begin{matrix} \mathbf{O} \\ \\ \mathbf{I} \end{matrix} \end{array} \right] \begin{bmatrix} \mathbf{F} \\ \mathbf{U} \\ \mathbf{E} \end{bmatrix}$$

where Φ, Θ and Ψ denote respectively the covariance matrices of **F**, **U** and **E**. When in addition **F**, **U** and **E** are assumed to be mutually independent, and Φ, Θ and Ψ are unrestricted matrices, then, according to the theory in the section above, asymptotic robustness will hold for the estimates of the parameters of β and γ, for the vector of residuals and also for the alternative χ^2 goodness-of-fit statistics. This will also be the case when the matrix Θ and Ψ are block-diagonal and mutual independence among the components can be assumed due to their partition. In the case of asymptotic robustness, the usual NT estimate for Γ can be used to obtain standard errors of estimates, and to evaluate asymptotic distributions of the test statistics, for any estimation method used (for instance, ULS, ML, etc.). In particular, the conventional ML analysis will be correct and optimal even for non-normal data. Note however that the standard error of the estimates of the variances of factors, methods and errors (that is, the estimates of elements of matrices Φ, Θ and Ψ) can not be based on normal theory if such components are non-normally distributed. If there are restrictions on the matrices Φ, Θ and Ψ, then for robustness results to hold the random constituents with restricted variances and covariances should be normally distributed (or, alternatively, all the variances and covariances among these random constituents should be unrestricted). When variances are restricted, but an equivalent parameterization of the model exists in which variances and covariances of non-normal constituents of the model are unrestricted, then robustness will hold for the χ^2 goodness of fit statistic and vector of residuals but not for variances of parameter estimates.

The RMM model

Now we will consider repeated multitrait multimethod (RMM) data (Saris, 1982). There we have responses on different factors (F_{ij} =1, 2, ..., j), different methods (i=1, 2, ..., I) and at different points in time (t =1, 2, ..., T). We write the model as

$$R_{jit\alpha} = \lambda_{jit} T_{ij\alpha} + E_{jit\alpha}$$

where

$\alpha = 1, 2, \ldots\ldots, N$
$T_{ji\alpha} = a_{ji} + b_{ji}F_{j\alpha} + U_{ji\alpha}$

and a_{ji} and b_{ji} are coefficients of the model. Here T_{ji}, E_{jit}, F_j and U_{ji} are random variables. Combining the above two equations, we get

$R_{jit\alpha} = \lambda_{jit}a_{ji\alpha} + \lambda_{jit}b_{ji}F_{j\alpha} + \lambda_{jit}U_{ji} + E_{jit\alpha}$
or
$R_{jit\alpha} = \alpha_{jit}\, a_{ji} + \beta_{jit}\, F_{j\alpha} + \gamma_{jit}U_{ji\alpha} + E_{jit\alpha}$

with the restriction that the ratio $\beta_{jit} / \gamma_{jit}$ (= b_{ji}) does not change with t. Note that when $t \equiv 1$ and $U_{ji} \equiv U_{(j'i)}$, $j \neq j'$, we get the MTMM model. When the vector-variables **F**, **U** and **E** are introduced, the RMM model can be rewritten in matrix notation as:

$$\mathbf{R} = \mathbf{a} + \left[\begin{array}{c:c:c} \beta\text{'s} & \gamma\text{'s} & \begin{matrix}\mathbf{O}\\ \\ \mathbf{I}\end{matrix} \end{array}\right] \begin{bmatrix} \mathbf{F} \\ \mathbf{U} \\ \mathbf{E} \end{bmatrix}$$

with additional restrictions on the β's and γ's, say g(β's, γ's) = 0. We also assume a block diagonal structure for the matrix of covariances of (F', U' and E')'.

When the F's, U's and E's are mutually independent (and not only uncorrelated) and Φ, Θ (or diagonal Θ) and Ψ (or diagonal Ψ) are free parameters, one can base inferences for β's and γ's on the NT form for Γ. In particular, the standard ML approach gives correct and optimal inferences even with non-normal data.

However, care should be taken not to restrict variances or covariances of non-normal variables. Under such restrictions, if an equivalent parameterization of the model exists for which the non-normal variables have free variances and covariances, then the robustness results still apply for the χ^2 goodness-of-fit statistics and vector of residuals.

CONCLUSIONS

In structural equation modeling assumptions are needed in order

1. to guarantee that the estimates are consistent for the parameters of interest, and
2. to evaluate the precision of the estimates and significance level of test statistics.

With respect to the first aim, the typical type of analyses (ML and WLS) are robust against violation of distributional assumptions; i.e., estimates remain consistent for any type of WLS analysis and distribution of **z**. (It should be noted, however, that the first aim is sensitive to structural misspecification.) A typical assumption used for the second aim is that the vector **z** of observable variables follows a multivariate normal distribution.

In relation to the first aim, distributional misspecification may have consequences for efficiency, as well as the power of test statistics (see Satorra, 1989); that is, some estimation methods may be more precise than others for a given specific distribution of **z**. For instance, ADF-WLS is asymptotically optimal under a variety of distributions of **z**, while the asymptotic optimality of NT-WLS may be lost when the data is non-normal.

Violation of a distributional assumption may have consequences for the second aim. However, recent theory, such as described in this chapter, suggests that asymptotic variances of estimates and asymptotic null distributions of test statistics derived under the normality assumption may be correct even when **z** is non-normal, provided certain model conditions hold (the conditions (a) and (b) of theorem 1). That is, in a specific application where **z** is non-normally distributed, the assumption that **z** is normal is a "working device" that facilitates calculation of the correct distribution of statistics of interest. This corresponds to what has been called asymptotic robustness.

With respect to the MTMM and RMM models, we reach several conclusions. When consistent estimation of the parameters of the model is the main goal, then any of the many available estimation methods for continuous data will do: whatever the distribution of the observable variables, if the model is correctly specified then all the minimum discrepancy estimates are consistent. Of course, matters of efficiency remain to be considered; i.e., some estimation methods may be more precise than others.

When we further want to assess the sampling variability of the estimates, i.e. to evaluate standard errors, then their distributional considerations need to be taken into account. If the data is normally distributed then standard ML analysis will of course provide asymptotically correct approximations to the standard errors of the estimates; further, the ML estimates are optimal. Without the imposition of the normality assumption, the same can be said for the ADF-WLS analysis: it provides optimal estimates and (theoretically) correct approximations to their asymptotic standard errors; however, this ADF approach puts heavy reliance on asymptotic approximations and is very demanding in computations. For most typical sample and model sizes, ADF-WLS may fail. An alternative approach is to use any estimation method, possibly a non efficient one (e.g. ULS), and then obtain a correct evaluation of the standard error of the estimates by using the methods described in the section on the asymptotic distribution-free estimate of Γ (in this way fourth-order sample moments are used only in assessing sampling variability, which contrasts with the ADF-WLS analysis where fourth-order sample moments also play a role in computing the estimates). Fourth-order moments can be avoided completely if we use the approach given in the section on asymptotic robustness.

In that section we see that when we interchange uncorrelation and mutual independence, and allow the variances and covariances of non-normal constituents of the model to be unrestricted, then we can still evaluate sampling variability of some parameter estimates by acting as if the data were normally distributed. Now, however, two types of parameters need to be distinguished: the α's and β's on the one hand, and the variances and covariances of the factors, method and error effects on the other. Asymptotic robustness applies to the first type of parameters but not to the second; that is, using the NT estimate of Γ we can compute appropriate standard errors for the estimates of parameters α's and β's, but not for the estimates of the mentioned variances and covariances. In particular, the standard ML analysis for the α's and β's is correct and optimal. The same pattern of asymptotic results holds equally for the MTMM and the RMM models.

When evaluation of the model by means of a χ^2 goodness of fit statistic is the main concern, then distributional considerations are needed from the very beginning. Of course, when data is normally distributed then the conventional ML χ^2 goodness of fit statistic is adequate and optimal. The same can be said for the ADF-WLS without the requirement of a distributional assumption; however, this approach relies heavily on asymptotics and costly computations. An alternative to this ADF-WLS would be first to use another estimation method and then the associated χ^2 goodness of fit statistics based on a distribution-free consistent estimate of Γ; with this approach, the fourth-order sample moments are used only minimally. In order to test the model, fourth-order sample moments can be avoided completely if we use the asymptotic robustness results given in the section on asymptotic robustness of normal theory inferences.

When uncorrelation and independence are interchanged and the variances and covariances of nonnormal constituents of the model are let free (for the specified model, or for a reparameterized equivalent version of it) then, given any estimation method, we can construct a χ^2 goodness of fit statistic based simply on a NT consistent estimate of Γ; in particular, the conventional NT χ^2 goodness of fit statistics is in effect asymptotically χ^2 distributed when the model holds. The same type of results applies for the vector of residuals; to evaluate the covariance matrix of the vector of residuals we can simply use a NT estimate of Γ instead of the correct one based on fourth-order sample moments. As in the case of the evaluation of the standard errors of the estimates, in constructing χ^2 goodness of fit statistics asymptotic robustness results for MTMM are of the same type as for the RMM models.

In the above models care should be taken when correlation among non-normal constituents of the model is allowed. In that case, for asymptotic robustness to hold, the grouping of the non-normal constituents in blocks of variables which are mutually independent but have unrestricted variances and covariances within the block should be possible; if we want such asymptotic robustness results to apply at least to the goodness of fit statistics and vector of residuals, such a grouping should be possible for an equivalent reparameterization of the model.

In conclusion, it is important to compute sampling variability for estimates and test statistics using appropriate formulae, without requiring that the estimation procedure be the "best" in some sense. We have seen that such computations can be carried out correctly under the wrong assumptions with respect to the distribution of the vector of observable variables, provided some additional model conditions hold. Generally speaking, the accumulation of such additional model conditions strengthens the usual assumption of uncorrelatedness among some random constituents of the model to the assumption of stochastic independence.

REFERENCES

Aigner, D.J., & Goldberger, A.S. (eds.) (1977). *Latent variables in socio-economic models*. Amsterdam: North-Holland.

Anderson, T.W. (1984). Estimating linear statistical relationships. *Annals of Statistics, 12*, 1-45.

Anderson, T.W. (1987). Multivariate linear relations. In T. Pukkila and S. Tuntanen (Eds.), *Proceedings of the second international conference in statistics*. Tampere, Finland.

Anderson, T.W. (1989). Linear latent variable models and covariance structures. *Journal of Econometrics, 41,* 91-119.

Anderson, T.W., & Amemiya, Y. (1988). The asymptotic normal distribution of estimators in factor analysis under general conditions. *The Annals of Statistics, 16,* 759-771.

Andrews, F.M. (1984). Construct validity and error components of survey measures: A structural modeling approach. *Public Opinion Quarterly, 48,* 409-442.

Bentler, P.M. (1983). Some contributions to efficient statistics in structural models: specification and estimation of moment structures. *Psychometrika, 48,* 493-517.

Bentler, P.M. (1985). *Theory and implementation of EQS, a structural equations program*. Los Angeles: BMDP Statistical Software.

Bentler, P.M. (1986). Structural modeling and psychometrika: An historical perspective on growth and achievements. *Psychometrika, 51,* 35-51.

Bentler, P.M., & Dijkstra, T. (1985). Efficient estimation via linearization in structural models. In P.R. Krishnaiah (Ed.), *Multivariate analysis VI*. Amsterdam: North-Holland.

Browne, M.W. (1974). Generalized least square estimators in the analysis of covariance structures. *South African Statistical Journal, 8,* 1-24.

Browne, M.W. (1982). Covariance structures. In D.M. Hawkins (Ed.), *Topics in applied multivariate analysis*. Cambridge: Cambridge University Press.

Browne, M.W. (1984). Asymptotically distribution-free methods for the analysis of covariance structures. *British Journal of Mathematical and Statistical Psychology, 37,* 62-83.

Browne, M.W. (1987). Robustness of statistical inference in factor analysis and related models. *Biometrika, 74,* 375-384.

Browne, M.W., & Shapiro, A. (1988). Robustness of normal theory methods in the analysis of linear latent variable models. *British Journal of Mathematical and Statistical Psychology, 41,* 193-208.

Campbell, D.T., & Fiske, D.W. (1959). Convergent and discriminant validation by the multimethod-multitrait matrix. *Psychological Bulletin, 56,* 81-105.

Ferguson, T. (1958). A method of generating best asymptotically normal estimates with application to the estimation of bacterial densities. *Annals of Mathematical Statistics, 29*, 1046-1062.

Fuller, W. (1987). *Measurement error models*. New York: Wiley.

Goldberger, A.S., & Duncan, O.D. (1973). *Structural equation models in the social sciences*. New York: Seminar Press.

Jöreskog, K.G. (1977). Structural equation models in the social sciences. In P.R. Krishnaiah (Ed.), *Applications of statistics*. Amsterdam: North-Holland.

Jöreskog, K.G. (1981). Analysis of covariance structures. *Scandinavian Journal of Statistics, 8*, 65-92.

Jöreskog, K.G., & Sörbom, D. (1983). *LISREL user's guide*. Chicago: International Educational Services.

Lee, S.Y., & Jennrich, R.I. (1979). A study of algorithms for covariance structure analysis with specific comparisons using factor analysis. *Psychometrika, 44*, 99-113.

Mooijaart, A., & Satorra, A. (1987). *Robustness of ML and ADF methods in the analysis of covariance structures: Heteroskedastic models*. Paper presented at the 1987 annual meeting of the Psychometric Society.

Muirhead, R.J. (1982). *Aspects of multivariate statistical theory*. New York: John Wiley.

Muthén, B. (1987). *LISCOMP: Software for advanced analysis of L.S.E. with a comprehensive measurement model*. Mooresville, IN: Scientific Software.

Magnus, J.R., & Neudecker, H. (1986). Symmetry, 0-1 matrices and jacobians: A review. *Econometric Theory, 2*, 157-190.

Mooijaart, A., & Bentler, P.M. (1987). *Robustness of normal theory statistics in structural equation models: Technical report 87-11*. Leiden: Leiden University.

Rao, C.R. (1965). *Linear statistical inference and its applications*. New York: John Wiley.

Saris, W.E. (1980). Linear structural relationships. *Quality and Quantity, 14*, 205-225.

Saris, W.E., & Stronkhorst, L.H. (1984). *Causal modeling in nonexperimental research: An introduction to LISREL analysis*. Amsterdam: Sociometric Research Foundation.

Satorra, A. (1989). Alternative test criteria in covariance structure analysis: A unified approach. *Psychometrika, 54*, 31-51.

Satorra, A. (1989). *Robustness of inferences based on the normality assumption: the instrumental variables case*. Paper presented at the European meeting of the Psychometric Society, Leuven, Belgium.

Satorra, A., & Bentler, P.M. (1986). Some robustness issues of goodness of fit statistics in covariance structure analysis. *Proceedings of the Business and Economic Statistics Section*: 549-554.

Satorra, A., & Bentler, P.M. (1988a). *Scaling corrections for statistics in covariance structure analysis*. UCLA Statistics Series #2. Los Angeles: University of California.

Satorra, A., & Bentler, P.M. (1988b). *Model conditions for asymptotic robustness in the analysis of linear relations*. UCLA Statistics Series #4. Los Angeles: University of California.

Shapiro, A. (1985). Asymptotic equivalence of minimum discrepancy function estimators to G.L.S. estimators. *South African Statistical Journal, 19*, 73-81.

Shapiro, A. (1986). Asymptotic theory of overparameterized structural models. *Journal of the American Statistical Association, 81*, 142-149.

chapter 5

The choice of a model for evaluation of measurement instruments

W.E. SARIS

In the previous chapters the problems which occur when the MTMM approach is used in the evaluation of measurement instruments have been discussed. These problems led to a lengthy discussion of the advantages and disadvantages of the different models and their necessary restrictions. In this chapter we report on the results of this discussion. We first discuss the differences between the two approaches: the MTMM and the RMM approach. Memory effects and the technical problems mentioned earlier are then discussed. Finally a design for evaluation studies will be proposed.

DIFFERENCES BETWEEN THE MTMM AND RMM MODEL

In order to clarify the differences between the different models we will start with the presentation of a measurement model which contains the characteristics common to both models (figure 1). This is the model used by Alwin and Jackson (1979) and by Heise and Bohrnstedt (1970).

In this model a distinction is made between theoretical variables that we are interested in (F_i) but which are not observable, and observed variables (x_{ij}) and true scores (T_{ij}). The latter variables are equal to the observed variables corrected for random measurement error. Furthermore, latent variables M_1 and M_2 are introduced to indicate the covariance between the true scores, which is due to the specific methods used to measure the different variables. Finally, unique variance (u_{ij}), which is specific for specific true scores, is added to allow for the possibility that the true scores for the same factor differ not only by the variance due to the common method but also because of an interaction between the method and the specific item.

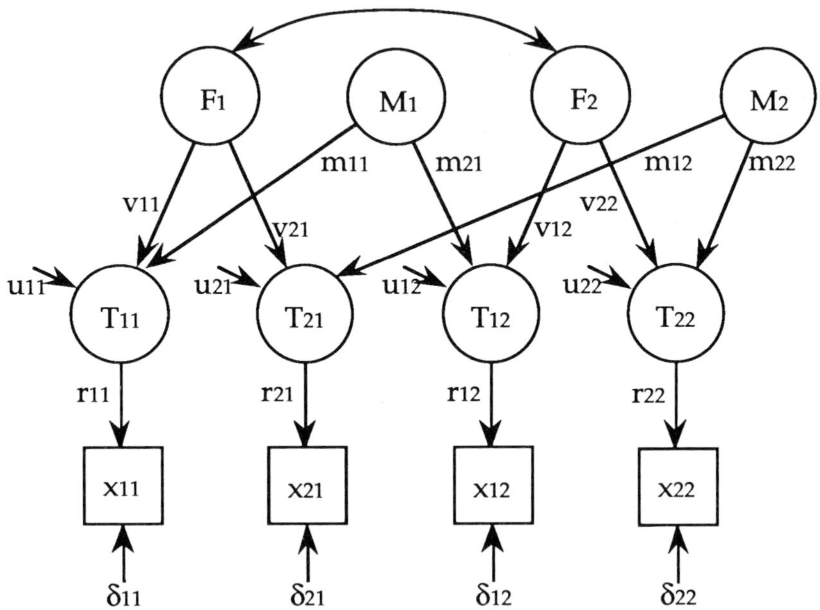

figure 1: A general measurement model

Following the definitions suggested by Saris in this volume, v_{ij} represents the validity of the jth variable measured with the ith method; $1\text{-}v_{ij}^2$ represents the invalidity of variable j measured with the ith method. A part of this invalidity is method specific variance which can be decomposed in method specific variance m_{ij}^2, where m_{ij} represents the method effect of method i on variable j, and the unique variance, var(u_{ij}), for the same variable. Finally, r_{ij}^2 is the reliability of variable j measured with the ith method.

Although these coefficients are not identified in this model, their meaning is clear. For their estimation at least two observations with each instrument are necessary and at least three different methods should be used for each variable. However, such an experiment is unpractical but the model is nevertheless useful to clarify the meaning of the different parameters.

If the RMM approach is used all the different coefficients which have been specified in the figure, except the method effects (m_{ij}), can be estimated. The estimation of these coefficients would require that at least three theoretical variables are measured in three different ways. We would then have to use a repeated MTMM model, which is unpractical because of the frequent repetition of the different questions (6 times for each different topic). However, if one of the designs which has been specified by Saris is used, estimates of all other parameters can be obtained as well as very detailed information about the size of the total systematic error, the validity and the reliability as defined above. An estimate of the systematic error or unique variance in the different measurement instruments is also obtained, which is especially attractive. This allows the possibility of determining whether the different methods really measure the same variable or not.

Although the information obtained by the RMM approach seems attractive as it gives the information which we need to evaluate measurement instruments one can also argue in favour of the coefficients obtained from the MTMM approach. In the MTMM approach the true scores are ignored because only one observation is made with each instrument. Consequently the distinction between u_{ij} and e_{ij} cannot be made. The coefficient, which is called the "validity coefficient" (λ_{fij}), is the product of $v_{ij}*r_{ij}$, the so-called "method effect" (λ_{mij}) is equal to $m_{ij}*r_{ij}$ and the so called "reliability" (R^2) is equal to $(v_{ij}*r_{ij})^2 + (m_{ij}*r_{ij})^2$.

Normally the MTMM design does not facilitate the estimation of the same coefficients as the RMM model. Only if there is no other source of systematic error beside the method effects can the coefficients of figure 1 be derived using the MTMM approach. In that case $R^2 = r_{ij}^2$ and on the basis of this information the other coefficients can also be obtained from the estimated parameters. A disadvantage of this approach is that one can not test this condition, which is not likely to be fulfilled. This does not mean, however, that the obtained information is irrelevant.

The results obtained suggest that the MTMM approach provides us with estimates of the effects of the latent variables on the observed variables, which are called the validity (λ_{fij}), the method effect (λ_{mij}) and the explained variance of the observed variables, which are called the reliability. Although these coefficients are not pure estimates of the characteristics of the measurement instruments, these parameters are certainly useful if relationships between different variables are studied. The validity and reliability coefficients are necessary to obtain consistent estimates of the correlations and effects between the theoretical variables. For example from figure 1 we get:

$$r(x_{11},x_{12}) = \lambda_{f11} * r(F_1,F_2) * \lambda_{f12} + \lambda_{m11} * \lambda_{m12}$$

and from this result follows:

$$r(F_1,F_2) = \frac{r(x_{11},x_{12}) - \lambda_{m11} * \lambda_{m12}}{\lambda_{f11} * \lambda_{f12}}$$

Thus in order to derive the correlation between the factors of interest we need estimates of the effects which can be obtained using the MTMM approach. The argument given above suggests that

- With the MTMM approach item specific variance and measurement error cannot be distinguished, and therefore one can not say whether the questions measure the same variable except for method specific variance.
- With the RMM approach the method effects cannot be detected, and therefore the inflation in the correlations between the observed variables, which is due to method correlation, cannot be corrected.

If the researcher ensures that exactly the same question is asked and only the method is varied, the item specific variance will probably be very small and can be ignored. However, the lack of information about the method effects seems to be a serious omission.

It would seem that for theoretical reasons, in order to determine whether questions really measure the same variable, the RMM approach is preferable. On the other hand, for practical purposes the coefficients of the MTMM approach, such as validity coefficients and method effects, are very important for deriving consistent estimates of the parameters of interest: the correlations between latent variables and the effects of latent variables on each other. For this reason the MTMM approach seems preferable.

Although this seems the proper conclusion it is important to stress its consequences: one has to use the same formulation of the question for all different methods. This rule has to be taken very literally. Let us apply it to satisfaction with health and examine what it means for three different methods:

How satisfied are you with your health?
Choose an answer from the following categories

1. very satisfied,
2. satisfied,

3. neither satisfied nor dissatisfied,
4. dissatisfied,
5. very dissatisfied

How satisfied are you with your health?
give a number between 1 and 10

1= very satisfied
10= very dissatisfied

These questions are the same and only the measurement scale is different. Even in such a case there would be some doubt as to whether the questions measure the same opinion, but this assumption can be accepted without a test. For evidence on this point we refer to Saris (1982). If the following question is used a test would certainly be necessary:

Some people worry about their health a lot. How about you - how worried are you about your health?

1. very worried,
2. quite worried
3. a little worried
4. not at all worried

This question introduces a new component: the emphasis of the question has shifted from "satisfaction" to "worry", which are not the same concepts. There can be considerable doubt about the equality of these three questions. Therefore, it is not surprising that Rodgers et al. (1988) found very low validity, high method effects and very high residual variance between these questions. This could be explained by item specific variation; in other words, the three questions do not measure the same variable.

This type of variation in questions should be avoided in the experiments. If there is any doubt about the possibility of item specific variance a RMM experiment should be performed to test for it or the whole MTMM questionnaire should be repeated later in a panel to see if a specific component exists. This valuable suggestion was made by Andrews on the basis of his recent research.

Repetition of questions and memory effect

A criticism of the RMM model is that it requires the repetition of various questions twice in one interview session. This procedure seemed questionable to some participants, given the possibility of memory effects. A simple experiment with a questionnaire showed that the same problems occur in the MTMM approach. This does not alleviate the problem in any way. In all cases the possibility of memory effects must be taken into account. So far it is not clear how this has been done in previous experiments. Its necessity can be illustrated by the study of Hox (1986) who asked questions on 5 different aspects of life satisfaction with 5 different methods in one mail questionnaire. Looking at the correlation matrix obtained we see that the correlations of variables measuring the same trait with different methods increases dramatically from around .5 in the beginning of the interview to around .8 at the end. It is not clear what this means but the most likely explanation is a memory effect combined with consciousness raising effects due to the repeated measurement.

In order to check this hypothesis it was decided that an experiment should be carried out to determine how many questions are necessary between repeated questions, to counteract such an effect.

SOME TECHNICAL PROBLEMS

Before a decision can be made in favour of the MTMM approach it should be clear that the design and model do not generate too many technical problems. These problems have also been discussed and an overview of the chosen solutions will be given.

Convergence problems

The MTMM model was heavily criticized for convergence problems when the data were analyzed with LISREL. In many instances the program could not estimate the parameters of the model. It was suggested by Költringer (1990) and Saris (1990) that this was a consequence of overfactoring. These remarks were in line with earlier remarks of Rindskopf (1984). Andrews, who did not have these problems in his study, suggested that this problem could be due to the lack of restrictions in the model. In his study he had introduced constraints on the method effects, specifying that they should be the same for all traits. This seems to be a plausible assumption. This suggestion was therefore evaluated on different data sets.

In data sets of Saris (1990) all MTMM models with the constraints mentioned above converged if the method factors were uncorrelated. For the same data the estimation did not converge in 5 out of 8 cases with the same constraints and unconstrained correlations between the method factors. Unacceptable solutions were found in the other three cases. The same problem has also been detected by Rodgers et al. (1988). When these correlations were restricted (fixing some at zero, leaving others free) mixed results were obtained: acceptable solutions were sometimes obtained, but frequently not. We will come back to the problem of correlated factors in one of the following sections.

Identification and design

Andrews also suggested that the problems caused by the repetition of questions could be reduced by using a 4 by 2 design, i.e. with four traits and two methods. This remark was however contrary to that of Költringer who held that at least three methods are required for identification. The 4 by 2 MTMM model without any constraints would not be identified. However, if the equality constraints are introduced on the method effects, as specified above, then the model would be identified. This does not imply that this 4 by 2 model should be used. In chapter 8 it has been shown that very different as well as irregular results can be obtained with different 4 by 2 models derived from a data collection using a 4 by 3 design. It was shown that the results are very unstable due to the fact that only two indicators per factor were used. This problem has also been mentioned by Boomsma (1983).

Memory effects and correlations between the method factors

During the conference Saris presented two correlation matrices for the same variables obtained from different groups, one of which had no knowledge about the topic and the other with some information. These matrices were completely different. In the matrix of the uninformed people high method effects could be seen and correlations between the substantive variables were low. In the matrix for the informed people the method effects were very small but the substantive correlations were much higher. This result had been replicated in other studies as well. In these studies correlated method effects were not necessary at all for informed people, while for the other topics about which the same respondents had little knowledge, correlations between line and number responses were needed.

It seems that the respondents answer questions differently if they have an opinion than when they are uninformed. In the latter case they try to be consistent between methods, not caring about what they say. This can explain the high method effects and high correlations between the methods in the uninformed case. Given these results it seems that for the time being topics about which the respondents have little knowledge should be avoided. It could perhaps be shown that these effects are smaller if the interval between the various repetitions of the questions is longer.

THE CHOSEN PROCEDURE

The discussion reported above has led to a number of decisions with respect to the approach to be used and these will be reported below.

Model choice and model test

A test to chose a model from the possible sets of models suggested by Költringer and Saris is necessary. It was shown that the χ^2 test which is normally used leads to many difficulties because of the varying power of the test in different data sets.

Therefore the general conclusion was that testing should be avoided as much as possible and one specific model should be fitted to all data sets. As we want to estimate the validity, invalidity and random measurement error, the best candidate is the model of Andrews which assumes the method effects are identical. This factor model is specified as follows:

$$Y = \Lambda F + E$$

where

- Y is a 12-1 vector of observed variables (four traits, each measured with three different methods)
- F is a 7-1 vector containing the four trait and three method factors
- E is a 12-1 vector of random error components
- Λ is a 12-7 matrix of effect parameters

The covariance matrix of the error components (E) of this model is specified as a diagonal matrix.

table 1: Specifications of the matrix Λ

	F1 (trait1)	F2 (trait2)	F3 (trait3)	F4 (trait4)	F5 (method1)	F6 (method2)	F7 (method3)
Y1	$\lambda_{1,1}$				$\lambda_{1,5}$		
Y2	$\lambda_{2,1}$					$\lambda_{2,6}$	
Y3	$\lambda_{3,1}$						$\lambda_{3,7}$
Y4		$\lambda_{4,2}$			$\lambda_{1,5}$		
Y5		$\lambda_{5,2}$				$\lambda_{2,6}$	
Y6		$\lambda_{6,2}$					$\lambda_{3,7}$
Y7			$\lambda_{7,3}$		$\lambda_{1,5}$		
Y8			$\lambda_{8,3}$			$\lambda_{2,6}$	
Y9			$\lambda_{9,3}$				$\lambda_{3,7}$
Y10				$\lambda_{10,4}$	$\lambda_{1,5}$		
Y11				$\lambda_{11,4}$		$\lambda_{2,6}$	
Y12				$\lambda_{12,4}$			$\lambda_{3,7}$

table 2: Specifications of the correlation matrix of the factors

	F1 (trait1)	F2 (trait2)	F3 (trait3)	F4 (trait4)	F5 (method1)	F6 (method2)	F7 (method3)
F1	1.0						
F2	$ps_{2,1}$	1.0					
F3	$ps_{3,1}$	$ps_{3,2}$	1.0				
F4	$ps_{4,1}$	$ps_{4,2}$	$ps_{4,3}$	1.0			
F5	0	0	0	0	1.0		
F6	0	0	0	0	0	1.0	
F7	0	0	0	0	0	0	1.0

If the respondents have an opinion on the question topic and the time between the repetitions is not too short, this model will be quite close to the true model for the MTMM experiments. Incidentally, correlations between the factors may be needed. In that case these correlations are introduced one after the other while combinations of correlations are avoided if they are not absolutely necessary. The reason for this is that the introduction of correlations has an effect on the estimates of validity and invalidity.

Therefore it should be ascertained which correlations have to be introduced and which left out.

The evaluation of the model should be done on the basis of the reduction in χ^2 (taking into account the power of the test), the size of the estimated correlation coefficient and the lack of alternative possibilities of correcting the model. In order to introduce a correlation between the methods a very large improvement in fit and a large coefficient are needed. In addition, there should be no other possible correlation between the method factors. A large correlation coefficient has a value of .4 or larger. A smaller value has no effect at all on the model and the fit.

With these rules most data sets will lead to an acceptable model. Occasionally two correlations will be needed but one should then seriously consider whether there are other problems in the data collection such as too short intervals between a repetition or lack of knowledge on the part of the respondents.

Estimation and standard errors

Satorra (1990) clearly indicated that there is no problem with respect to estimation in case the normality assumption is not satisfied. The estimated values of the parameters are always consistent. The χ^2 test and the standard errors might be wrong in that case but if we assume the independence of the errors from the factors instead of the the uncorrelatedness these statistics are correct then even if the ML estimators are used. In general, this alternative assumption is reasonable, but an exception will be indicated below. The use of the asymptotic distribution free (ADF) methods leads to more problems than solutions and should, therefore, be avoided.

There is one case which requires further consideration: the situation where categorical variables with few categories are used. Here the relationships between the factors and the observed variables become nonlinear and the errors are not independent of the factors. This case requires special attention. The general consensus was that in this case the polychoric or polyserial correlations should not be used as substitutes for the normal product moment correlations because of the assumption of normality which is once again introduced. For the time being it is perhaps the best to avoid categorical scales with fewer than 5 categories (to mention an arbitrary cutting point).

Data cleaning

In the meetings a plea was made for cleaning of the data before the analyses are done, as outliers can affect the results considerably.The following steps should be carried out:

1. Look at scattergrams or tables of different methods for each trait. If outliers are found in the corners of the table or plot, list the scores of these respondents.
2. Look at the scores. Are these mistakes systematic errors which can be corrected or not? Systematic errors are reversals of the scales or due to use of too short a scale (a scale of 100 points instead of 1000 points for example). Such errors can be corrected and the case can be saved. Incidental errors like forgetting a zero (100 instead of 1000) can also be corrected if other judgments clearly indicate that this error has occurred.
   If no correction rule can be specified than the value should be made missing.
3. Look at the scattergram and tables again to see whether the picture has improved. If not, repeat the same procedure.

When making such corrections, one should of course be very careful not to improve one method while neglecting another. The number of corrections for the different methods should be registered for later analysis.

The procedure for the data analysis has been specified on the basis of this discussion. There are two issues which will be decided later: problems with category scales with few categories, and the number of cases needed for the study. On both points further studies will be done.

REFERENCES

Alwin, D.F., & Jackson, D.J. (1979). Measurement models for response errors in surveys: Issues and applications. In K.F. Schuessler (Ed.), *Sociological methodology 1980*. San Francisco: Jossey-Bass.

Boomsma, A. (1983). *On the robustness of LISREL (maximum likelihood estimation) against small sample size and non-normality.* Amsterdam: Sociometric Research Foundation.

Heise, D.R., & Bohrnstedt, G.W. (1970). Validity, invalidity, and reliability. In E.F. Borgatta and G.W. Bohrnstedt (Eds.), *Sociological methodology 1970*. San Francisco: Jossey-Bass.

Hox, J.J.C.M. (1986). *Het gebruik van hulptheorieën bij operationalisering, een studie rond het begrip welbevinden*. Amsterdam: Universiteit van Amsterdam.

Költringer, R. (1990). Analysis of mutitrait multimethod matrices. In W.E. Saris and A. van Meurs (Eds.), *Evaluation of measurement instruments by meta-analysis of multitrait multimethod studies*. Amsterdam: North-Holland.

Rindskopf, D. (1984). Structural equation models: empirical identification, Heywood cases, and related problems. *Sociological Methods and Research, 13*, 109-119.

Rodgers, W.L., Herzog, A.R., & Andrews, F.M. (1988). Interviewing older adults: Validity of self-reports of satisfaction. *Psychology and Aging, 3*, 264-272.

Saris, W.E. (1982). Different questions, different variables. In C. Fornell (Ed.), *A second generation of multivariate analysis: Vol. 2. Measurement and evaluation*. New York: Preager.

Saris, W.E. (1990). Models for evaluation of measurement instruments. In W.E. Saris and A. van Meurs (Eds.), *Evaluation of measurement instruments by meta-analysis of multitrait multimethod studies*. Amsterdam: North-Holland.

Saris, W.E. (1990). The choice of a research design for MTMM studies. In W.E. Saris and A. van Meurs (Eds.), *Evaluation of measurement instruments by meta-analysis of multitrait multimethod studies*. Amsterdam: North-Holland.

Satorra, A. (1990). Robustness issues in the analysis of MTMM and RMM models. In W.E. Saris and A. van Meurs (Eds.), *Evaluation of measurement instruments by meta-analysis of multitrait multimethod studies*. Amsterdam: North-Holland.

Part 2

Design of the multitrait multimethod studies

Introduction

In Part 1 we have seen that the multitrait multimethod approach has many difficulties. Nevertheless, this model has been chosen because of the information which can be obtained with it. Consequently, in the studies repeated observations will have to be made for the same items using different methods. We anticipate that this will lead to problems. In the MTMM model it is assumed that the different observed variables for the same traits do not affect each other. Memory effects have thus been excluded as a possibility.

The validity of this assumption was the focus of a recent study by van Meurs and Saris. It was found that there is good evidence that these memory effects exist and that they affect the results of the MTMM analysis. We shall now report on this research and discuss the consequences for research designs which might eliminate these effects or at least reduce them.

We also discuss the effect of the choice of the traits on the results of the analysis of MTMM models. The possibility cannot be excluded that the specific choice of traits has an effect on the resulting quality coefficients. In an experiment by van Meurs this point has been tested and effects on the estimates were found.

At the end of Part 2 we discuss the designs which should be used in order to reduce these effects as much as possible. Although these two studies were not reported at the conference in Amsterdam but at the conference in Brussels in 1989, we have taken the liberty presenting them here because of their importance for the approach which is discussed in this book.

chapter 6

Memory effects in MTMM studies

A. VAN MEURS AND W.E. SARIS

The MTMM approach to evaluate methods of data collection in attitude and opinion research is based on the repetition of questions on the same item with different methods. Such an approach allows estimates of validity, invalidity and reliability coefficients, which are very important in correcting for measurement error in substantive research. However, repeating questions about the same item can also have its drawbacks. Many respondents will try to be consistent in their answers across methods. They will of course succeed if they can remember their previous answers. Such memory effects can bias the estimates of the quality indicators for measurement instruments in an MTMM model. In this paper we will discuss the following topics:

- the possible effects of memory
- to what degree respondents are able to remember their previous answers
- the factors which affect memory
- the consequences for the design of MTMM studies.

The data discussed in this paper were collected in April 1989. The respondents (N=1537) of the NIPO-telepanel (de Pijper and Saris, 1986; Verwey, et al., 1989) answered two questionnaires with an interval of two weeks between their presentation. Three questions in the first wave concerned the attractiveness of the three main Dutch political parties measured using a 10-point category scale. The same response scale was used for the attitude towards six public services and three life satisfaction questions. After a number of other questions half of the sample (N=795) was asked if they could remember their initial answers to items 1 to 3 and to give their opinion again on items 4 - 6 using the line production

The authors would like to thank H. Seip for his contribution, and H. Hartman and K. de Rooy for their work on the survey design.

method. The other half of the sample (N=751) was tested for their memory of the questions 4-6 and they were asked to express their opinion about the items 1-3 using the line production method. Two weeks later all of the respondents cooperating (1381) completed a second interview. Again two groups answered a memory question on three items (group 1 on items 7-9, group 2 on items 10-12). Furthermore, all respondents gave categorical judgements on all items and line productions for all questions not evaluated in that way. The complete design is summarized in figure 1. The chosen design allows the correlations and MTMM estimates to be obtained for all variables with a short interval between the observations (within the same interview) and with a two week interval between the observations.

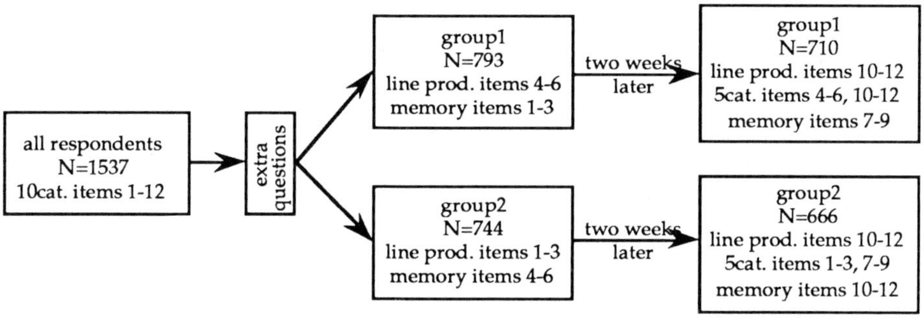

figure 1: Design of the study

THE POSSIBLE EFFECTS OF MEMORY

Repeating the same questions in a single survey can have its drawbacks. Some respondents are annoyed because repeating the same questions is considered as a waste of time. Others feel that they are not trusted, and that their answers are secretly checked by the researchers. These problems can be minimized by introducing the repeated questions with some sort of explanation. Our experience is that respondents then answer the questions seriously.

A more troublesome problem arises when respondents can remember their previous answers. If respondents are recalling their previous answers in the formulation of their reply to the repeated questions, then so called memory effects which are assumed to be zero in the MTMM model, might arise. If these memory effects are ignored, the estimates of the quality of the measurement instruments will be biased. This will generally lead to overestimation of the quality of the instruments, especially their validity.

Another possibility is that repetition of a question will give the respondents the impression that the questions are of some importance to the researcher and, consequently, they will think more carefully about their answers the second time. The answer could then change between the first and second time, thereby leading to lower correlations and lower estimates of the quality of the measurement instruments. This problem has been discussed in a set of papers by Jagodzinski et al. (1987) and Saris and van den Putte (1988) as the so-called "socratic effects" of asking questions.

From this brief discussion it is not clear what the expected effect will be in practice for the size of the correlations and the MTMM parameters. In the experiment described we could estimate the different effects on the correlation and the parameter estimates of the MTMM model after an interval of approximately 9 minutes (on average) within an interview and after an interval of two weeks in two different waves of the same panel. The results for the correlations are summarized in table 1.

table 1: Correlations between two methods with different time intervals

	methods:	10 cat-line		5 cat-line	
	time interval:	±9 min.	2 weeks	±9 min.	2 weeks
1	PvdA (labour)	.851	-	-	.824
2	CDA (chr. dem.)	.863	-	-	.830
3	VVD (liberals)	.799	-	-	.733
4	health care	.825	-	-	.499
5	post delivery	.760	-	-	.513
6	public transport	.820	-	-	.614
7	police	-	.605	.736	-
8	social security	-	.547	.698	-
9	road system	-	.550	.731	-
10	income	-	.713	.791	-
11	housing	-	.741	.762	-
12	gen. living situation	-	.603	.692	-
	mean correlation:	.820	.627	.735	.669

This table shows very clearly that, except for the first three items, the correlations between two measures of the same variable after 9 minutes is much higher than after two weeks. This effect is also clear for the mean correlations. We can see from the means that the items which had higher correlations after 9 minutes have lower correlations after 2 weeks. Although the topic might have some effect (for the first three items) these results definitely suggest that the memory effect is very strong and it seems most likely that all estimates will be biased upwards.

In table 2 the estimates of the MTMM models for the same items have been given. The solid line in the table indicates when the two week interval occurred for a specific item. The two week interval occurred for the first 6 topics after the second observation. For items 7-12 the two week interval occurred after the first observation of the variable.

table 2: MTMM parameter estimates for validity and reliability

	validity			reliability		
	10cat	line	5cat	10cat	line	5cat
1 PvdA (labour)	.881	.926 |	.919	.884	.906 |	.859
2 CDA (chr. dem.)	.858	.939 |	.912	.900	.900 |	.873
3 VVD (liberals)	.855	.904 |	.837	.892	.844 |	.765
4 health care	.946	.873 |	.575	.915	.854 |	.416
5 post delivery	.895	.852 |	.598	.820	.817 |	.469
6 public transport	.920	.889 |	.685	.866	.893 |	.563
7 police	.668 |	.868	.867	.632 |	.880	.787
8 social security	.657 |	.820	.855	.596 |	.790	.765
9 road system	.652 |	.810	.888	.607 |	.813	.813
10 income	.744 |	.895	.900	.719 |	.869	.835
11 housing	.808 |	.891	.856	.813 |	.877	.770
12 gen. living situation	.691 |	.844	.816	.684 |	.790	.699

| = a two week interval between observations

This table indicates that the instruments which were employed together in one interview have much higher validity and reliability estimates (except for the first three topics) than the items which have been asked separately, irrespective of whether this observation occurred before or after the interview with the two observa-

table 3: Mean parameter estimates for validity and reliability (derived from table 2)

	validity			reliability		
	10cat	line	5cat	10cat	line	5cat
items 1-6:	.893	.898 I	.754	.880	.869 I	.658
items 7-12:	.703 I	.870	.864	.675 I	.837	.778

I = a two week interval between observations

tions for these variables. In table 3 the means of these coefficients are given and this result is here even clearer.These results suggest that the memory effect leads to an overestimate of the quality measures of the variables which have been observed in the same interview. In this data set we did not find any indication of reduced correlation due to so-called "socratic effects".

Given this conclusion and the size of the effects of the time difference on the estimated coefficients, we will concentrate for the rest of this chapter on the memory effects and try to establish first how many people remember their answers correctly.

MEMORY OF PREVIOUS ANSWERS

In order to determine whether a respondent could remember his or her previously given answers the following question was asked:

"Can you remember exactly what you answered on the question about .... ?

This question measuring the self indication by the respondents was followed by a question which asked for the exact response. If the respondents claimed that they remembered the answer, we asked:

"What number was given by you? "

If the respondents said that they did not remember the answer the following question was asked:

"Estimate the number which was given by you:"

The first question gives us an indication of how many people think that they can correctly remember the answers. The last two questions provide us with an estimate of how many people can exactly reproduce their answers. We are purposely not saying that they remember the answer correctly because it is quite likely that some respondents will give a correct answer by chance, even thought they no longer recall the previous answer. In tables 4 and 5 the results of the analysis of these three questions are summarized.

table 4: Self-indication of memory

	memory %	no memory %	no memory, no answer %	
±9 minute interval:				
1 PvdA (labour)	72.0	22.5	5.5	N=793
2 CDA (chr. dem.)	73.6	21.4	4.9	N=793
3 VVD (liberals)	70.9	22.7	6.4	N=793
4 health care	79.6	18.0	2.4	N=744
5 post delivery	84.9	12.6	2.4	N=744
6 public transport	77.9	20.1	2.0	N=744
two week interval:				
7 police	18.0	70.3	11.7	N=710
8 social security	16.8	68.7	14.5	N=710
9 road-system	15.3	66.9	17.7	N=710
10 income	27.8	61.6	10.7	N=666
11 housing	38.9	55.4	5.7	N=666
12 gen. living situation	27.9	63.8	8.3	N=666

We distinguish between the question repeated within the same interview and in two different interviews. Within the same interview the time between the first question and the memory question on the same item was on average 9 minutes. For the first 6 topics the memory question was asked within the same interview. For the last 6 topics the respondent had to go back in memory more than 2 weeks.

In table 4 we see that on average 76% of the respondents claimed to remember exactly what they had said previously in the same interview whereas 24% did not remember it. Of all respondents 4% said that it was impossible to give an estimate of their previous answer on these first six topics.

For the topics 7 to 12 we have comparable results only for the two week interval. Approximately 24% said that they could remember their answer exactly and the remainder said that they could not remember their answer at all. For both sets of answers these percentages do not have to indicate how many people indeed remember their previous answer correctly. In order to check this assumption we asked further extra questions about the given number. In table 5 the results of these questions are summarized.

table 5: Percentage of correct repreduction of previous answers

			self-indication*		
		memory		no memory	
±9 minute interval:					
1 PvdA (labour)	57.1	66.2	N=571	33.9	N=222
2 CDA (chr. dem.)	57.9	66.4	N=584	34.1	N=209
3 VVD (liberals)	59.9	68.5	N=562	39.1	N=231
4 health care	67.0	73.4	N=592	42.1	N=152
5 post delivery	74.1	79.3	N=632	44.6	N=112
6 public transport	72.1	79.1	N=579	47.3	N=165
two week interval:					
7 police	40.6	57.0	N=128	37.1	N=582
8 social security	33.1	42.9	N=119	31.1	N=591
9 road system	32.8	52.3	N=109	29.2	N=601
10 income	41.1	53.5	N=185	36.3	N=475
11 housing	48.7	63.7	N=259	39.2	N=401
12 gen. living situation	47.7	54.8	N=186	44.9	N=474

* all group means are significantly different at a .05 prob. level, two tail T-test

The results in table 5 clearly indicate that not all respondents who claimed to remember their previous answer could reproduce it. Only approximately 70% of this group could reproduce the correct number. However approximately 36% produced the correct answer even though they said that they did not remember their previous answer. It should be noted that for these respondents it did not matter whether the question was asked within the same interview or after two weeks, in the second wave of the panel study. Also, the respondents who after two weeks claimed that they could remember the previous answer did not perform any better on the memory task. In this group the percentage of correct answers was close to 36% as well.

These results suggest that 36% of respondents can produce the same answer to a repeated question even though they have forgotten their previous answer. This phenomenon is not surprising because most respondents are not likely to have changed their opinion in such a short time period. Therefore, there is a relatively high probability producing the same answer. Given our results we estimate the probability that they produce the same answer again without any memory effect at 36%. It follows that of the respondents who can reproduce the answers correctly after 9 minutes approximately (70%-36% or) 34% is probably due to memory effects while the others reproduce the result because of the stability of their opinion.

It would be interesting to know how high this percentage is after 4, 8, 12, 15 or even 20 minutes. If there is a considerable decay in memory after some time, this information would be useful in the design of MTMM studies. This aspect can be studied because a computer-assisted data collection procedure was used in which the length of the interview differed, depending on the branches of the questionnaire the different respondents had to answer: for some the time between when the question was first asked and the memory question was less than 3 minutes while for others the interval was more than 20 minutes. If we split up the sample into different groups according to the time between the question and the memory questions and recalculate the percentages of correct answers for the different topics we get the results presented in table 6, which are graphically represented in figure 2.

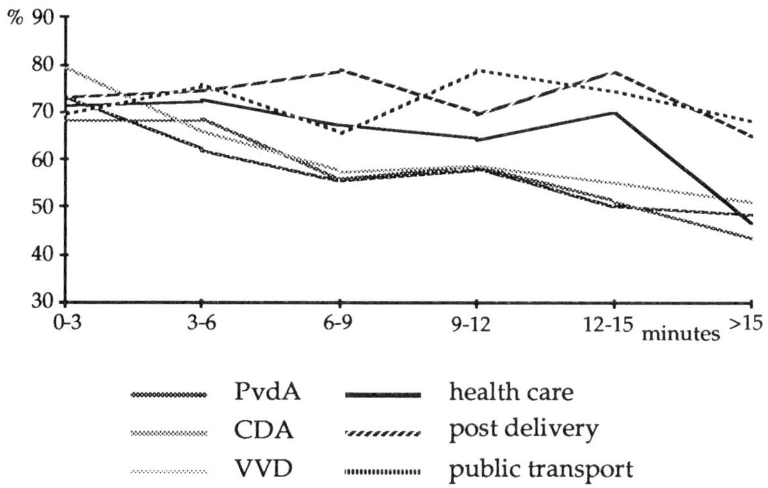

figure 2: The percentage of correct answers for different time intervals

table 6: The percentage correct answers for different time intervals (in minutes)

	0-3	3-6	6-9	9-12	12-15	15-20	> 20
1 PvdA (labour)	73.0	61.9	55.5	58.1	50.0	48.6	46.2
2 CDA (chr. dem.)	68.3	68.5	55.9	58.8	51.2	43.1	43.6
3 VVD (liberals)	79.4	65.8	57.3	58.8	54.9	52.8	46.2
4 health care	71.4	72.7	67.0	64.0	69.9	40.9	52.8
5 post delivery	73.2	74.6	78.8	69.4	78.5	61.4	69.4
6 public transport	69.6	75.8	65.5	78.9	74.2	63.6	72.2

In this table and figure we see that for the political topics 1-3 there is a strong reduction in the number of correct answers to the memory question as the time interval increases from 3 minutes to more than 20 minutes. The reduction in correct answers is around 25%. However for the topics 4 - 7, this did not occur.

These results suggest that the time interval between the repeated questions is not the only factor involved. In order to get a better understanding of the effects of different factors on the percentages of correctly reproduced answers we have made a further analysis which will be reported in the next section.

FACTORS WHICH DETERMINE SHORT TERM MEMORY

We hypothesized that the following factors play an important role in the recall of previous answers:
- the time interval between questions
- the type of questions presented between repetitions
- the extremeness of the response to the question when first presented

The first factor has been discussed above. The second factor concerns the type of questions between the first presentation of a question and its repetition. In our earlier experiments we found that respondents are unable to remember an answer to an earlier question if a long series of similar questions has been asked before its repetition. For example, in a judgement study we asked respondents to evaluate 25 occupations with respect to status (Saris, 1982). After these 25 occupations were evaluated by line production, the same question was asked and the respondents were expressed their opinions by magnitude estimation. In this and similar experiments it was clear that the respondents did not remember their answer to a specific item due to the large number of similar judgments they were asked to give.

In the present experiment the intervening questions were of a political nature, asking about party identification, probabilities to voting for a party, etc. We believe that it is due to this type of question during the time interval that the political questions show more decay in memory than the other questions.

The third factor is very evident. We hypothesize that it is much easier to remember an answer which is represented by an endpoint of a scale than one on any other point of the scale. Not only are the answers easier to remember but the respondent probably also has an extreme opinion which is likely to be quite salient and central for him or her (Schuman and Presser, 1981; Billiet et al., 1985).

To test these hypotheses we have studied the effect of the first factor, the time interval between the two questions (T), on the percentage of correct answers to the memory questions (C) for four different situations which are determined by the other two factors. According to our hypotheses we expect that:

- there will hardly be any effect due to the time interval if subjects have given an extreme response
- the effects will be stronger for the questions on political issues than for those on the social services.

The dependent variable in the analysis was the percentage of correct answers in the different sugroups, determined by three factors :

- time interval
- political topic or not
- extreme response or not

In this analysis a simple linear regression model was tested using the multiple group analysis of LISREL (Jöreskog and Sörbom, 1988) where the different categories are weighted by the number of people in the subgroup. A model was hypothesized where the regression coefficients are zero for the cases with extreme responses. The results are presented in table 7, which shows that this model fits to the data well. Further restrictions could not be introduced because they led to rejection of the model at the 5% level.

In this table we see that for the two conditions where respondents have given extreme answers the effect of the time intervals is zero. There is no effect at all due to the length of time between when the question was first asked and the memory questions. This result is independent of the topic of the questions. On average the

table 7: The effect of the time interval (T) on the correct reproduction of the previous answer (C) under different conditions.

topic	type of responses to the first question: no extreme values	extreme values
political	C = 59.0 -.94 T	C = 80.6 + 0.0 T
not political	C = 75.4 - .50 T	C = 71.4 + 0.0 T

$\chi 2$ = .43 df = 2 prob= .8

respondents correctly reproduce the answer they have given earlier in 75% of the cases. Assuming that in 36% of the cases the correct answer is due to stability of opinion, we concluded that for a large number of respondents (40%) a memory effect can be expected to occur.

Where no extreme answers have been given the effect of the time interval (T) depends on the topic of the repeated questions:

- if the questions between the first question and its memory question are of the same type (political), then every extra minute in the time interval will reduce the percentage of correct answers by approximately 1% (on average).
- if the the first question is of a different type than the intervening questions (non-political) the effect of the time interval is only half as large: an increase of two minutes reduces the number of correct answers by 1%. In addition, the mean percentage of correct answers is also much higher. Respondents thus have a better recall of their previous answer whatever the time interval.

CONCLUSIONS

In this chapter we have shown first of all that the effect due to the interval between the repeated observations on the size of the correlation coefficients, and consequently on the estimates of the MTMM model parameters, can be considerable.

Secondly our analysis has indicated that these increases in correlations and estimates of the parameters are probably due to the large number of respondents (34%) who can reproduce their responses to previous questions after a short period of time. We also found that in approximately 36% of the cases the answers were correctly reproduced even though the respondents said that they could not remember these answers (after a period of two weeks).

These correct answers are not due to a memory effect but to the stability of opinion over time.

Thirdly, it has been shown that the possibility of remembering the previous answers is not only a function of time. Two other factors played an important role: extreme responses to the questions when first presented, and the type of questions asked between their repetitions.

In this study respondents who gave extreme answers were very capable of reproducing their previous answers and this capability was not affected by longer time intervals between the presentation of the questions within one interview and by the type of questions asked in the interval.

These factors did have an effect when respondents did not give extreme scale points to the questions of interest. If we can accept the results obtained and are willing to extrapolate over time and generalize for other topics, which is of course very questionable, then the following practical conclusions could be drawn:

1. In a situation where similar questions are asked between two repeated observations of the same variable, the memory effects will disappear if the time interval between the observations is at least 15 minutes. Approximately 30 to 35 questions can be asked in such a time interval. This result is in agreement with our experience with judgement tasks with more than 25 similar items (Saris, 1982).
2. In a situation where questions of a different type have been asked between the repeated observations a time interval of 80 minutes can eliminate the memory effect completely.

These results pose serious problems for the MTMM approach to evaluate the quality of measurement instruments. It should be clear from the RMM approach suggested by Saris in this volume that any procedure based on repeated observations would also have similar problems. The memory problems are actually even more serious than the discussion here has indicated as the MTMM and RMM approaches require more than 2 observations for each variable.

These problems require further consideration in the design of such evaluation studies. A few remarks can be made immediately:

1. If the study is done in one wave one should use a very long interview with many questions of the same kind concerning the same topic between the repeated observations.
2. It is worth checking whether only two methods, and therefore only two repetitions of the same topic, can be used instead of three. This would reduce the problems considerably.

3. Panel studies using one method per wave and at least two weeks between the waves should be considered as a possible design. It would be necessary however to check for changes in opinion between the waves. This can be done by a quasi experimental design where under one condition exactly the same questions are used at three points in time and the Quasi Simplex model (Wiley and Wiley, 1970) is used for the test of the change in opinion.

The results presented in this chapter suggest that if these recommendations are not taken into account in an MTMM study it is most likely that memory effects will play an important role and that all validity and reliability estimates will be subsequently overestimated. Our guess is that that this has occurred in most MTMM studies to date.

REFERENCES

Billiet, J., Loosveldt, G., & Waterplas, L. (1984). *Het survey-interview onderzocht. Effecten van het ontwerp en gebruik van vragenlijsten op de kwaliteit van de antwoorden*. Leuven: Sociologisch Instituut.

Jagodzinski, W., Kühnel, S.M., & Schmidt, P. (1987). Is there a "socratic effect" in nonexperimental panel studies? *Sociological Methods and Research, 15,* 259-302.

Jöreskog K.G., & Sörbom, D. (1988). *LISREL 7: A guide to the program and applications*. Chicago: SPSS.

Pijper, W.M. de, & Saris, W.E. (1986). *The formulation of interviews using the program INTERV*. Amsterdam: Sociometric Research Foundation.

Saris, W.E. (1982). Different questions, different variables. In C. Fornell (Ed.), *A second generation of multivariate analysis: Vol. 2. Measurement and evaluation*. New York: Preager.

Saris W.E., & Putte, B. van den (1988). True score or factor models: a secondary analysis of the Allbus test-retest data. *Sociological Methods and Research, 17,* 123-157.

Schuman, H., & Presser, S. (1981). *Questions and answers in attitude surveys*. London: Academic press.

Verwey, M., Saris, W.E., Mosselman, K., & van Doorn, L. (1989). Tele-interviewing in practice: Household incomes show large monthly fluctuations. *Marketing and Research Today, 17,* 230-240.

Wiley, D.E., & Wiley, J.A. (1970). The estimation of measurement error in panel data. *American Sociological Review, 35,* 112-117.

chapter 7

Trait combinations in the MTMM model

A. VAN MEURS

INTRODUCTION

Questions used in public opinion research can be investigated for differences in validity, reliability and method effects using a multitrait multimethod (MTMM) model. The measurement qualities of different methods for a number of topics can be compared using a set of structural relationships defined in LISREL.

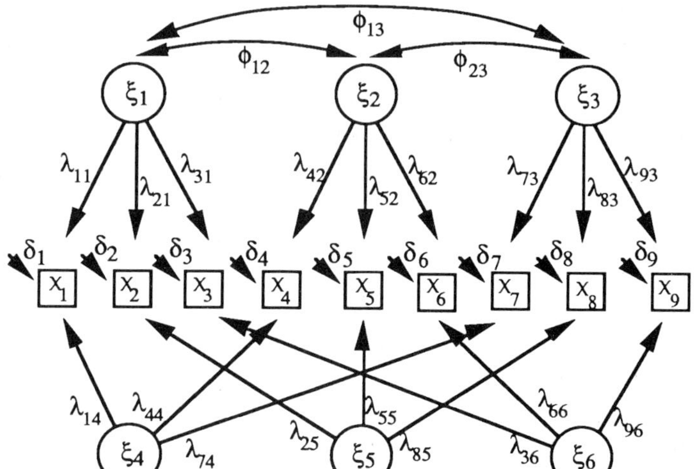

ϕ_{ij} = covariance $\xi_i\ \xi_j$ $\qquad \lambda_{14} = \lambda_{44} = \lambda_{74}$
ϕ_{ii} = variance $\xi_i = 1$ $\qquad \lambda_{25} = \lambda_{55} = \lambda_{85}$
Θ_δ = variance δ_i $\qquad \lambda_{36} = \lambda_{66} = \lambda_{96}$

figure 1: MTMM33 model

The author would like to thank W.E. Saris for his continuous help and support.

This MTMM model is based upon the repeated measurement of a set of traits with a variation in method, for example different response scales. Such a model with four items and three different observations is called an MTMM43 model. Figure 1 shows the path diagram of an MTMM33 model.

Andrews (1984) used an MTMM model similar to that in figure 1 to obtain parameter estimates for the validity, reliability and method effects for all the methods and topics of a very large number of previously obtained datasets. In a meta-analysis these parameter estimates were explained by the different characteristics of each specific combination of method and trait.

In the International Project on Evaluation of Measurement Instruments a similar approach will be adopted, but with the advantage that the survey design can be established beforehand, including all variation in different methods and traits, in all possible combinations. In this chapter I would like to concentrate on one aspect of the MTMM model that could be important when choosing a set of items and in planning the future meta-analysis.

To compare all MTMM parameters in one meta-analysis, every dataset has to be analyzed with the same LISREL model. Following the approach of Andrews, the International Project on Evaluation of Measurement Instruments decided upon the following specifications and constraints for the MTMM model (Saris, 1990):

- The variation of the latent variables is standardized at 1.
- The λ_{ij} of each method for the same trait are constrained to be equal. This assumption is not always appropriate and necessary, but having free λ_{ij} will sometimes result in estimation problems, and therefore hinder a comparison of parameters from all datasets.
- The matrix Θ_δ contains only diagonal elements, the model is assumed to have no correlated errors.
- The *validity* of a measure x_i of a trait ξ_j is the magnitude of the direct structural relation between ξ_j and x_i.
- The *method effect* of a method ξ_k on a measure x_i is the magnitude of the direct structural relation between ξ_k and x_i.
- The *reliability* of x_i is the total effect (R^2) that all variables (except δ's) have on x_i.

The standard model has a set of four traits and three methods, but MTMM models with three traits can also be used to derive coefficients that indicate the measurement quality of traits and method combinations, in all kind of different datasets. In the meta-analysis the validity and reliability of each x_i will be explained by all possible characteristics of the particular trait and method that have direct structural relations with x_i in the MTMM model. The validity and reliability of the survey question x_6 in figure 1 will be explained using the characteristics of the trait ξ_2, and the method ξ_6 with which it was observed.

This chapter considers one aspect of the MTMM model that has not been discussed yet: traits and methods do not operate individually in the model, but in combination with other traits and methods. If, for example, ξ_1 and ξ_3 affect the direct structural relation between ξ_2 and x_6, a complete meta-analysis will be extremely complicated.

In this chapter an experiment with trait combinations will be discussed, to investigate the possibility of the effect of different combinations of traits on the validity and reliability of x_i. A similar effect due to method combinations is theoretically possible as well. A brief discussion on this point will be presented in the next chapter.

POSSIBLE EFFECTS OF TRAIT COMBINATIONS

Every possible trait combination in an MTMM model will inevitably have a different correlation matrix. The validity and reliability estimates for one particular trait should however be the same in every model. The parameter estimates that express the measurement quality of each trait should be stable, and not be influenced by the specific combination of traits in the MTMM model.

It would be interesting to know how robust the MTMM estimates are against changes in trait combinations. In fact, we are only interested in the extent changes in the trait combination will affect the λ and δ parameters representing the validity, reliability and the method effects.

The ϕ parameters in the model are very likely to change with every alteration in trait combination, because these parameters express the relationship between the traits. The Φ matrix, however, should not have any effect on these parameters. Possible differences in the λ and δ estimates for different trait combinations can be caused by various factors:

1. In LISREL the parameters are estimated using a full information estimation procedure, so all parameters could theoretically be affected by a change in the correlation matrix.

2. In our model the relationships between all δ-variables expressed by the $\theta\delta$ matrix are fixed at 0 (except for the diagonals). In reality these constraints will not always be justifiable. If two x-variables do have correlated errors, and their common $\theta\delta$ parameter is fixed, the inter-correlation of this pair of x-variables will affect the estimated λ parameters functioning as indirect effects between these two x-variables. A change in trait combinations could therefore act upon the λ parameters representing the method effects, and possibly the validity coefficients, although this is less likely .
3. If the method effects or validity coefficients are affected by trait combinations, the reliability coefficients will be influenced as well. By definition the reliability of x_i is the sum of the squared direct relations that all variables (except δ's) have on x_i. These direct relations can only be method effects and validity coefficients.
4. Some traits could have a non-normal distribution, or have non-linear relationships with other traits. These problems, specific to one individual trait or trait combinations, will result in correlated errors and a bad fit of the model. Because the parameters are estimated in LISREL by a full information estimation procedure, the parameter estimates will differ in value and accuracy for every change in fit of the model.

THE EXPERIMENT

To determine the effect of different trait combinations on the MTMM parameters, the same dataset discussed in chapter 6 was used. The twelve different traits were measured on three different response scales: a five- and ten-point scale, and line production. The number of possible combinations between 3 traits and 3 methods was limited to 40, each containing a unique combination of three traits. Every single trait could be placed in 10 different MTMM models.

Estimates for the housing-parameters in combination with other traits

First, the results of this experiment will be explained by focussing on one variable, in this case the evaluation of the respondent's satisfaction with his or her own housing on a ten-point scale (item 11, 10cat scale). This measurement is placed in an MTMM model as a single variable, together with the measurement of the same item using two other response scales. These variables represented the three observations for the housing item. Together with three different observations of two other items the MTMM33 model is complete. The MTMM parameters for the housing trait were estimated together with two other personal traits: income and general living situation. The MTMM estimates for these three traits are shown in table 1.

table 1: MTMM results for 3 items: χ^2 = 102.15, df = 21, N=710

	validity			method effects			reliability		
	10cat	line	5cat	10cat	line	5cat	10cat	line	5cat
10 income	.744	.895	.900				.719	.869	.835
11 housing	.808	.891	.856	.368	.277	.212	.813	.877	.770
12 gen. living sit.	.691	.844	.816				.684	.790	.699

The LISREL estimation of the validity of the measurement of the housing item using a ten-point response scale is .808 in this model, and the reliability is .813. The method effect of the 10 category response scale in this model is set as equal for the three traits, and is estimated to be .368.

To investigate whether these three parameter estimates of item 11 in this combination are the same in other trait combinations, item 11 was placed in 9 other MTMM33 models. The ten different trait combinations are listed in table 2. This table gives a summary of the results of these ten different models for our example: item 11 measured on a 10cat scale.

table 2: The parameter values for item 11 (housing) measured on a 10-point category scale in 10 different MTMM models

item combinations	validity	method effect	reliability
4, 5, 11	.818	.094	.676
4, 6, 11	.816	.109	.673
4, 10, 11	.817	.155	.694
4, 11, 12	.824	.206	.729
5, 6, 11	.818	.120	.679
5, 10, 11	.818	.209	.717
5, 11, 12	.825	.278	.766
6, 10, 11	.820	.138	.694
6, 11, 12	.827	.202	.730
10, 11, 12	.808	.368	.813
mean *(st.dev.)*:	.819 *(.005)*	.188 *(.080)*	.717 *(.045)*

Table 2 shows that there are some differences in parameter estimates between the models. The estimates for the validity of the housing measurement on a 10-category response scale do not differ much. The standard deviation indicates that the estimates for the validity coefficients were fairly constant in all ten models.

The ten estimates for the method effect show a lot more variance. In each case, this parameter was estimated under the constraint that its value was the same for all three items. Therefore it is not surprising that this parameter changes when there are different traits in the model. The equality constraint of the λ's links the method effect of one trait measurement with the same method effect of other traits. A change in traits will affect all linked λ's if the method effects are different from trait to trait.

If the method effects show a lot of variance, the reliability coefficients should vary as well, because the reliability is a funtion of the method effects and the validity.[1] Indeed, the estimates for the reliability of the housing measurement on a 10-category response scale differ substantially. The minimum estimation is .673, and the maximum is .813. The finding that the reliability coefficients for this item differ in their first position after the decimal is an important but inconvenient one. For this item the assumption that the reliability is determined only by the characteristics of its own trait and method can no longer hold. This will complicate the forthcoming meta-analysis.

Results for all variables

As for the housing example in the previous section, twelve traits were estimated in an MTMM33 model. Each trait was combined with two other traits in ten different models. Because of a split half design in the questionnaire which was included for other purposes, not all combinations were possible. The items 1-3 and 7-9 were collected for one subsample (N=666), and the other half of the respondents answered the questions 4-6 and 10-12 using all three methods (N=710).

The next three tables give a summary of the distribution of the most important parameter estimates: the coefficients for validity, method effects and reliability.

[1] The total variance of a measure is the sum of the squared validity and unreliability coefficients, and the squared method effect (given linear relationships, a perfect estimation of the model and independence of these factors):

$$var_y = val^2 + meth^2 + unrel \Rightarrow (1\text{-}unrel) = rel = val^2 + meth^2$$

table 3: Distribution of validity coefficients for ten different trait combinations

		10cat	*stddev*	line	*stddev*	5cat	*stddev*
1	PvdA (labour)	.896	*.008*	.932	*.003*	.895	*.011*
2	CDA (chr. dem.)	.889	*.015*	.937	*.020*	.893	*.010*
3	VVD (liberals)	.880	*.014*	.896	*.005*	.824	*.010*
7	police	.691	*.009*	.870	*.009*	.849	*.008*
8	social security	.663	*.008*	.831	*.008*	.844	*.009*
9	road system	.654	*.008*	.831	*.010*	.879	*.009*
4	health care	.931	*.011*	.887	*.008*	.578	*.003*
5	post delivery	.881	*.011*	.863	*.005*	.603	*.002*
6	public transport	.909	*.007*	.905	*.007*	.681	*.002*
10	income	.794	*.018*	.882	*.011*	.894	*.006*
11	housing	.819	*.005*	.898	*.004*	.846	*.004*
12	gen. living situation	.714	*.010*	.847	*.007*	.802	*.006*

table 4: Distribution of the reliability coefficients for ten different trait combinations

		10cat	*stddev*	line	*stddev*	5cat	*stddev*
1	PvdA (labour)	.876	*.020*	.902	*.014*	.813	*.021*
2	CDA (chr. dem.)	.884	*.024*	.894	*.005*	.818	*.028*
3	VVD (liberals)	.866	*.028*	.832	*.021*	.711	*.027*
7	police	.545	*.033*	.799	*.037*	.735	*.021*
8	social security	.499	*.038*	.727	*.030*	.729	*.020*
9	road system	.486	*.046*	.738	*.032*	.782	*.017*
4	health care	.891	*.012*	.830	*.015*	.379	*.018*
5	post delivery	.813	*.011*	.795	*.015*	.407	*.028*
6	public transport	.851	*.009*	.867	*.015*	.506	*.026*
10	income	.677	*.020*	.824	*.024*	.831	*.013*
11	housing	.717	*.045*	.857	*.014*	.750	*.013*
12	gen. living situation	.569	*.053*	.768	*.010*	.694	*.013*

The tables 3, 4 and 5 resemble the summarized results of the housing example. The validity parameters (table 3) differ in their third, and sometimes second position after the decimal point. The amount of variance is not alarming however. The variation of the method effects throughout the experiment is rather large. We cannot say anything about the differences in estimates for each single trait-method combination because of the equality constraint.

table 5: Distribution of the method effects for ten different trait combinations

	mean	*stddev*
10 category response scale	.211	*.066*
line production	.173	*.060*
5 category response scale	.194	*.051*

In the standard MTMM model the λx_{ij} of the method effects are constrained to be equal.

The standard deviations of the parameter values for the reliability coefficients indicate that the estimates have a lot of variation in the second position after the decimal point, and even variation in the first position is no exception. The estimates for reliability coefficients for each trait are therefore very unstable, and depend on the trait combination in the MTMM model.

POSSIBLE EXPLANATIONS

In survey research, traits can influence each other. The combination of traits in questionnaires is just one of the factors determining the validity and reliability of a question. In an MTMM the trait combination model used to estimate the qualities should not have any influence. However, the LISREL estimates of the standard MTMM model that are important for the meta-analysis, change for individual traits if the selection of traits is varied. Ideally the parameter estimates should be robust for these alterations, but the method effects and reliability parameters in particular show a large variation across the models with different trait combinations. This variation could be caused by a number of factors.

The variation of the estimates that occurs when the trait combinations are changed could be an artefact of the specifications of the MTMM model itself. The model does not include structural relations to represent correlated error between the δ-variables, because having such parameters and method effects together will prevent the model from being identified. Most of the variation in the model related to this correlated error will find its way through the method effects. These λ's are likely to change for different trait combinations, because the correlated error will also be different. This is probably the cause of some of the variation in the method effects and the reliability coefficients in this experiment.

There is another aspect of the structural relations that will produce effects of trait combinations on the parameters: the constraint of equal values for the method effects of each trait. Because of this constraint the estimation of λ_{ik} is not only an expression of the relationship between one particular pair ξ_k and x_i, but is also affected by the structural relationships of this ξ_k with all other x-variables. In this way trait combinations can affect the λ_{ik}. This parameter in turn will affect the reliability, because the reliability of x_i is defined as the sum of the squared direct relations that all variables (except δ's) have on x_i. The equality constraint of equal values for the method effects of each trait links all the traits in the model with the method effects and reliability coefficients in the MTMM model. A change in trait combination could therefore have an effect on these parameters.

The equality constraint on the λ's causes an unwanted link between the traits in the model. The constraint is not always necessary for estimating the model. Setting the λ's free would allow discrimination between effects of the same method for different traits. If for instance one item is particularly salient for all respondents, and another is not, the method effects would differ between these two traits.

To investigate whether independent estimation of the λ's could reduce the effects of trait combinations on the parameter estimates, an additional experiment was done with the same data discussed earlier in this chapter. The 40 MTMM models were estimated without the constraint of equal values for the method effects of each trait. This was only possible for 14 models. Due to convergence problems in 15 cases the method effects had to be constrained for one trait in order to estimate the model, in 8 cases for two traits, and in three cases the equality constraint could not be removed at all.

The results (given in the appendix) were surprising. The method effects show differences for each item, even for the same method, but each method effect also has a very large variation for every different trait combination. Table 6 shows the mean standard deviation of the validity and reliability estimates for the models with and without the λ-constraint.

Removing the constraint did not result in smaller standard deviations for the reliability coefficients. On the contrary, without a constraint on the λ's there was much more variation between the reliability coefficients for each trait combination.

table 6: Mean standard deviation of parameter estimations for 40 models with different trait combinations

	10cat	line	5cat
std.dev. for validity:			
with λ-constraint	.010	.008	.008
without λ-constraint	.012	.008	.018
std.dev. for reliability:			
with λ-constraint	.028	.019	.020
without λ-constraint	.117	.072	.079

Some models in our experiment had a poor fit. If we assume that the structural model is correct, this poor fit indicates that other assumptions, such as distributional, are not fulfilled. A bad fit can result in unstable estimates, and a large unexplained variance. These problems will vary for every trait and trait combination. In this experiment, the questions concerning the political parties (items 1-3) suffered especially from bad fit in the model due to the U-shaped distribution of the responses. The fit of these traits increased when combined with other traits, but caused a worse fit for traits 7-9. The change in fit would probably have influenced the parameter estimates and the explained variance of the MTMM models. It is difficult to say which combination of traits is optimal.

CONCLUSIONS

This study has shown that the MTMM estimates for the quality of measurement instruments are not robust against alterations in the combination of traits. Different trait combinations have very little effect on the size of the validity coefficients but a considerable effect on the method effects and reliability coefficients. But if the validity coefficients are relatively stable then the variation in results is due to the variation in size of the residuals for the variables which are measured by the same method. Since we assume that the method effects are the same for each method and are independent of the variable measured, the residuals for the variables measured by the same method should also be the same if the model is correct.

There are two reasons why this situation will not occur in substantive research. The first is that the model may be misspecified, for example for some substantive variables the residuals are larger than for others. Secondly, the differences could occur because of sampling fluctuations.

It is very difficult to imagine what kind of misspecification would cause the residuals to be larger for some variables than for others. We are thus inclined to think in terms of sampling fluctuations. The reader should bear in mind that the method effects are mostly between .2 and .3 and subsequently the correlation produced by these method effects is between .04 and .09. However, for a sample of 500 respondents the correlation coefficient could vary, due to sampling fluctuations, from -.088 to +.088 if the population value is zero. Therefore it is very likely that the coefficients vary considerably for the different trait combinations. Increasing the sample size would not help very much because for a sample of 1000 cases the 95% confidence interval is still between -.062 and +.062.

Given this situation, introducing the constraints on the method effects would seem to be the most reasonable approach. The method effects will then be estimated in some way on the mean residual correlation between the variables measured using the same method. This alternative approach to stabilizing the method effects seems to be the most viable way to reduce the excessive sampling fluctuation in the estimates of the method effects and the reliability coefficients. In table 6 this point has been illustrated very clearly.

This study, however, has also shown that we should be cautious in our expectations concerning the explanation of the method effects because a large part of the variation in these coefficients is due to sampling fluctuation even if constraints are introduced in the models.

REFERENCES

Andrews, F.M. (1984). Construct validity and error components of survey measures: A structural modeling approach. *Public Opinion Quarterly, 48,* 409-442.

Saris, W.E. (1990). The choice of a model for evaluation of measurement instruments. In W.E. Saris and A. van Meurs (Eds.), *Evaluation of measurement instruments by meta-analysis of multitrait multimethod studies.* Amsterdam: North-Holland.

APPENDIX:
MTMM experiment for models without the λ-constraint

table 7: Validity coefficients for models with different trait combinations, and without λ-constraint

		10cat	*stddev*	line	*stddev*	5cat	*stddev*
1	PvdA (labour)	.894	*.010*	.932	*.004*	.896	*.011*
2	CDA (chr. dem.)	.887	*.018*	.939	*.004*	.889	*.010*
3	VVD (liberals)	.834	*.022*	.901	*.013*	.843	*.032*
7	police	.684	*.015*	.867	*.016*	.852	*.009*
8	social security	.662	*.005*	.827	*.011*	.852	*.014*
9	road system	.655	*.007*	.825	*.016*	.887	*.007*
4	health care	.931	*.013*	.859	*.005*	.578	*.003*
5	post delivery	.887	*.006*	.863	*.005*	.603	*.002*
6	public transport	.912	*.006*	.901	*.008*	.683	*.002*
10	income	.793	*.017*	.886	*.005*	.895	*.006*
11	housing	.815	*.005*	.900	*.004*	.846	*.004*
12	gen. living situation	.704	*.018*	.852	*.009*	.802	*.008*

table 8: Method effects for models with different trait combinations, and without λ-constraint

		10cat	*stddev*	line	*stddev*	5cat	*stddev*
1	PvdA (labour)	.300	*.176*	.122	*.065*	.110	*.072*
2	CDA (chr. dem.)	.189	*.132*	.165	*.187*	.141	*.079*
3	VVD (liberals)	.213	*.139*	.181	*.150*	.209	*.191*
7	police	.526	*.392*	.231	*.194*	.220	*.141*
8	social security	.362	*.260*	.148	*.102*	.258	*.136*
9	road system	.206	*.124*	.116	*.091*	.307	*.214*
4	health care	.149	*.095*	.264	*.163*	.233	*.104*
5	post delivery	.184	*.194*	.228	*.110*	.285	*.127*
6	public transport	.129	*.082*	.257	*.198*	.234	*.125*
10	income	.204	*.113*	.148	*.053*	.170	*.101*
11	housing	.201	*.144*	.211	*.162*	.187	*.064*
12	gen. living situation	.280	*.190*	.270	*.158*	.263	*.209*

table 9: Reliability coefficients for models with different trait combinations, and without λ-constraint

		10cat	*stddev*	line	*stddev*	5cat	*stddev*
1	PvdA (labour)	.835	*.101*	.894	*.017*	.810	*.024*
2	CDA (chr. dem.)	.874	*.052*	.903	*.030*	.848	*.088*
3	VVD (liberals)	.855	*.056*	.884	*.114*	.746	*.080*
7	police	.913	*.500*	.813	*.059*	.806	*.121*
8	social security	.637	*.211*	.762	*.067*	.755	*.055*
9	road system	.486	*.060*	.823	*.175*	.797	*.017*
4	health care	.898	*.039*	.843	*.036*	.426	*.141*
5	post delivery	.861	*.128*	.788	*.080*	.424	*.056*
6	public transport	.854	*.030*	.885	*.062*	.563	*.133*
10	income	.693	*.050*	.821	*.039*	.823	*.017*
11	housing	.731	*.061*	.855	*.020*	.777	*.107*
12	gen. living situation	.617	*.121*	.839	*.165*	.739	*.104*

chapter 8

The choice of a research design for MTMM studies

W.E. SARIS

In the first part of this book we have indicated a preference for the MTMM approach to evaluate the quality of measurement instruments. The choice of this model does not mean that the design of these studies is completely fixed. The number of traits and methods to be used have not been determined, nor have decisions been made concerning the number of questions between the repeated observations, the order of the questions, the kind of data collection techniques and research designs, and the minimal sample size.

During the meeting of the research group in Brussels two serious problems were raised with respect to the MTMM approach. First of all, van Meurs showed that the results, especially with respect to the method effects, depend heavily on the traits which are combined in the model. He showed experimentally that the variation in method effects is considerable for the different trait combinations. This result suggests that it will be difficult to explain the variation in method effects in the meta-analysis.

Van Meurs and Saris showed that the sizes of the validity coefficients differ considerably depending on the time interval between the different observations. Variables measured with three different methods, twice in one interview and once two weeks later in another interview, had considerably different validity coefficients. The former two had validity coefficients of .8 and the latter of .6. They also showed that in this specific case the decay in memory was very small. Further research led to the conclusion that variable time intervals are required, the length of which depends on the type of questions between the repetitions to prevent memory effects from occuring.

Given these complications, we believe that further study of the research designs is necessary. A question which immediately comes to mind is: Would it not be possible to do an MTMM analysis with 2 methods instead of 3. In order to answer this question I have looked at available data once more to see what happens if this simplification is introduced.

TWO VERSUS THREE METHODS

For five olympic sports respondents were asked how much they watched the television coverage of these events. Furthermore they were asked their evaluation of the programs. The respondents gave answers on a 5-point category scale, a 1000-point magnitude scale and a line production scale with a length of 38 positions. The sample consisted originally of 450 respondents and was randomly drawn from the Dutch population. There were 150 respondents who had watched all the events and therefore had scores for all variables. We analyzed the data of these respondents.

The data (15 variables) were firstly analyzed using an MTMM model with 5 factors and 3 methods with method effects constrained to be equal for the different variables measured by the same method. We then analyzed a part of the data (10 variables) with an MTMM model with 5 traits and only 2 method factors. Finally, we analyzed the same data while ignoring the method effects.

The results are summarized in table 1. This table shows that:

1. The results are very different if we use 3 instead of 2 methods.
2. Ignoring the method effects in the models with 15 variables does not have much effect on the estimated values of the parameters but has a strong effect on the fit of the model.
3. Ignoring the method effects in the models with 10 variables (2 methods) can give quite different results. In one case even a nonconvergence was encountered.

Let me briefly comment on these results. For the first three variables measured with categories, number and lines the correlations are .85 .87 and .92. The validity coefficients for the three methods were .9, .95, and .98. As can easily be checked, these coefficients reproduce the correlations quite well. Now imagine that we have only the first correlation. Many solutions are then possible for the validity coefficients. In the model with only two methods the estimated values are 1.0 and .87. These coefficients also reproduce the correlation reasonably well. However they are completely contradictory to the correlations for the three-method model. As the correlation between x_2 and x_3 is .92, the validity for x_3 has to be larger than 1 and the correlation between x_2 and x_3 should have been close to one and not .87.

table 1: Validity and method effects for different models

	5-3 model	5-2 models			5-0 models		
	var1,2,3	var1,2	var1,3	var2,3	var1,2,3	vars1,3	vars2,3
VALIDITY:							
categories	.897	1.00	.970		.895	.848	
	.901	.969	.982		.879	.921	
	.898	.840	.828		.854	.747	
	.945	1.04	1.01		.889	.991	
	.841	.929	.919		.832	.864	
magn. estim.	.949	.870		.959	.945		.949
	.937	.820		.986	.943		.972
	.942	.944		1.09	.953		1.05
	.969	.791		.988	.969		.975
	.927	.835		.953	.931		.948
line production	.975		.918	.958	.976	1.03	.971
	.962		.856	.927	.959	.917	.934
	.971		.983	.851	.965	1.11	.872
	.950		.808	.945	.967	.866	.962
	.968		.884	.944	.965	.929	.947
METHOD EFFECT:							
categories	.275	.000	.000				
magn. estim.	.180	.339		.053			
line production	.081		.296	.184			
PHI:							
	.425	.430	.431	.433	.429	.402	.436
	.132	.098	.123	.079	.132	.166	.083
	.293	.226	.216	.297	.297	.258	.319
	.327	.233	.278	.333	.331	.344	.342
	.411	.364	.375	.419	.411	.411	.421
	.039	-.074	-.096	.140	.041	-.01	.115
	.353	.362	.377	.343	.360	.363	.357
	.808	.832	.834	.793	.816	.868	.808
	.265	.180	.147	.300	.273	.214	.301
	.450	.425	.454	.443	.457	.487	.451
FIT:							
χ^2	182	50.6	41.6	71.2	334	127	123
df	77	23	23	23	80	25	25
prob.	.000	.001	.01	.000	.000	.000	.000

The results for the other solutions are very similar. They suggest that there is too much freedom in the two-method model. Because of some characteristic of the data a solution is found which is impossible in the three-method case.

Due to this freedom the validity coefficients can vary by more than .15 for the same data but different pairs of methods. This variation is not caused by the method effects because the same problems arise if the method effects are ignored.

The conclusion is therefore that three different methods are necessary as otherwise too much variation in the validity coefficients will occur. This can not be explained by method characteristics and is due to sampling fluctuation. This brings us back to the problem of designing the data collection in order to avoid memory effects as much as possible.

DESIGNS FOR MTMM STUDIES IN CROSSSECTIONAL RESEARCH

Given that three different methods are needed for each variable and that respondents can remember their previous answers quite well, we have to do something special to disturb their memory in order to avoid memory effects. I would like to make two suggestions.

In the first chapter in this part we have shown that the decay in memory appears to be much faster when monotonous judgement tasks are given to the respondents. This result does not stand alone. In the past I have done several methodological studies where I have asked respondents to give 25 judgements of different items on the same scale. Examples are evaluations of the status of occupations, the lengths of lines or attractiveness of incomes etc. After so many similar judgements the respondents did not remember their previous answers and we could repeat the same question with another instruction.

The second possibility is, of course, to increase the time interval between the repeated observations of the same variable. This would mean that these MTMM studies could only be performed in very long interviews. Our analysis has shown that time intervals between the repetitions of 15 minutes are required for monotonous judgement tasks and intervals of more than 30 to 40 minutes where there is a lot of variation in the questions (van Meurs and Saris, 1990).

A problem of this approach, besides the length of the interview, is that a possible effect due to the difference in method cannot be separated from a possible effect due to the ordering of the methods in the questionnaire. In order to study the last possibility the methods should be presented experimentally. With three different conditions all three different positions for three different methods can be specified. In each subsample sufficient cases should be used in order to determine whether the solutions are identical or not. If they are identical the data can be pooled. However, if the results are significantly distinct the analysis should be done on the separate samples and the order of the methods in the questionnaire should be introduced as one of the explanatory variables.

In addition there is the problem that respondents may change their opinion because of the questions which are asked about the topic. This phenomenon is not unusual in survey research. It can even be shown that these effects can be generated on purpose in cross-sectional research (Saris and Hartman 1990). It is thus necessary to test for opinion change. In order to do so another experimental condition can be added to the experiment whereby a fourth sample of the same population is presented the same questions about the same topics at the same places where the other repondents are presented three times with a different question on the same topic. Such data can be analysed with a Quasi Simplex model (Wiley and Wiley 1970) to test for change in opinion during the interview. If such a change has occurred its size can be determined and the estimation of the MTMM model can be adjusted by taking into account the estimated change in opinion over time. In that case a combination of a Quasi Simplex model and an MTMM model should be used with constraints on the parameters of the model. This is not the place to elaborate on this model but there is no doubt that this model would be identified and can even be tested. But the researcher should take care that the samples for each experimental condition are sufficiently large and drawn in the same way from the same population.

Summarizing this approach, I would recommend a design with four different experimental conditions in the data collection:
- group 1: in sequence methods 1, 2, 3
- group 2: in sequence methods 2, 3, 1
- group 3: in sequence methods 3, 1, 2
- group 4: the same method three times

The first three conditions will indicate if order has an effect. The fourth condition can be used to test if the opinion being tested has changed over time.

It will be clear from this discussion that this approach is far from ideal, given the length of the interview which is required. It can only be applied in substantive face to face interviews which take a lot of time (more than one hour) and where the repeated questions can be spread in different orderings through the questionnaire: one set at the very beginning, the second set in the middle and the third set at the end.

Another possibility is that these questions are subjected to methodological research where the reseacher can control all the questions. In such circumstances one can approximate the previously described situation of our methodological judgement studies with very monotonous questions between the repeated question.

Even though this design can be used for MTMM studies we have seen that many problems have to be overcome in order to use this approach well. In addition it does not allow for MTMM studies by telephone or tele-interviewing because of the length of the interview, and its use in mail surveys is doubtful. If such long interviews are used it is quite likely that the response will be very low and that the approach is impossible in practice. Given these problems and limitations we would also like to propose a design for MTMM studies in panel research.

DESIGNS OF MTMM STUDIES IN PANEL RESEARCH

In the first study mentioned in part 2 of this volume, we discovered that after two weeks nearly all respondents had forgotten the answers they had given to the questions in the previous interview. The respondents could therefore be asked the same questions again after two weeks using a different response method without risking memory effects. This approach also has the advantage that the interviews do not have to be as long as those in cross-sectional research. The disadvantage is, of course, that the respondents might have changed their opinion in the two week interval (Saris and Van den Putte, 1988). This risk is even larger when the time interval between the observations is longer.

However, his problem can be tackled by the same method we have suggested before. One can introduce an experimental design with four different conditions where in three conditions the order of the questions is varied while in the fourth the same questions are asked three times in order to determine whether the respondents have changed their opinion. This approach has been described above.

The application of this approach in panel research allows for MTMM studies using face to face interviews (even though that will be expensive), telephone interviews, tele-interviews and mail questionnaires. All possible data collection techniques can in principle be used for MTMM studies so the data collection techniques themselves can be compared.

SAMPLE SIZE

A final point with respect to the design of the MTMM studies is the size of the samples. In our studies with samples of 400 cases the validity coefficients, which are between .8 and .9, had standard errors around .04, and for the method effects, which are .25 to .3, the standard errors were around .02. This precision is acceptable for the further analysis which has to be done on these estimated values. On the basis of these results it seems sufficient to do MTMM studies with samples of 400 persons or more in the total sample.

If a design that includes four different subgroups is used the sample size would be reduced to 100 respondents in each subgroup. In that case we would expect the standard errors to increase by a factor 2. The standard error for the validity coefficient would then increase to .08 and to .04 for the method effect. Both seem too large for the design of a test with an accepetable power for differences in parameter values between the subgroups. For samples of 200 cases the standard errors would be approximately .06 and .03, which is already better but may still be too large. Therefore we suggest that the experimental designs include samples larger than 200 respondents for each subgroup. In that case the total sample for the pooled data, if the samples are not significant different, will be larger than 600 cases. Such a sample would certainly produce enough precision for the meta-analysis. If the subgroup parameters are significantly different there is enough precision for further analysis.

REFERENCES

Jagodzinski, W., Kühnel, S.M., & Schmidt, P. (1987). Is there a "socratic effect" in nonexperimental panel studies? *Sociological Methods and Research, 15,* 259-302.

Jöreskog K.G., & Sörbom, D. (1988). *LISREL 7: A guide to the program and applications.* Chicago: SPSS.

Meurs, A. van, & Saris, W.E. (1990). Memory effects in MTMM studies. In W.E. Saris and A. van Meurs (Eds.), *Evaluation of measurement instruments by meta-analysis of multitrait multimethod studies*. Amsterdam: North-Holland.

Saris, W.E., & Hartman, H. (in press). Common factors can always be found, but can they be rejected? *Quality and Quantity*.

Saris W.E., & Putte, B. van den (1988). True score or factor models: a secondary analysis of the Allbus test-retest data. *Sociological Methods and Research, 17*, 123-157.

Wiley, D.E., & Wiley, J.A. (1970). The estimation of measurement error in panel data. *American Sociological Review, 35*, 112-117.

Part 3

Meta-analysis

Introduction

In the third part of this book we will discuss the design of the meta-analysis which has been suggested by Andrews and discussed during the conference. With the term meta-analysis we mean the analysis which will be done using the results of the different studies performed in the different countries. For each measure in each study the "validity", "invalidity" and random error variance will be estimated. Beside that each measurement instrument can be characterized by a set of choices of the survey design characteristics. Therefore, if the data with respect to the measures quality and the design characteristics from all studies are brought together in one file the effects of the different design characteristics on the quality statistics can be studied and useful information for substantive research can be derived. Andrews has illustrated this procedure very clearly in his paper in Public Opinion Quarterly which has been reprinted in this volume.

During the conference especially two aspects of the meta-analysis have been discussed at length. The first point was the design of the meta-analysis, the second point was the choice of the characteristics of the measurement instruments which should be varied in the studies.

In this third part we will start with papers of Andrews and Mellenbergh on the design aspect. Then one paper about the choice of the measurement characteristics follows. Molenaar makes a comment on the choices of Andrews. After these contributions an extended summary of the discussion is given.

chapter 9

Some observations on meta-analysis of MTMM studies

F.M. ANDREWS

Although the results of analyzing different multitrait multimethod (MTMM) data sets are described in Andrews (1984, tables 4, 5, and 6), more detailed remarks about the approach and some special considerations relevant to it will be made in this contribution.

ON THE "CASES" FOR THE META-ANALYSIS

One needs to remember that a "case" in the meta-analysis consists of data for a particular trait, assessed using a particular method, for a particular group of people. In my analysis each case had the following information associated with it:

a. the measurement quality estimates for that case (i.e., estimates of validity, method effect, and random error)
b. the thirteen survey design/content features that characterize that measure and that are discussed in Andrews (1984)
c. the type of respondent for whom the measurement quality estimates were computed.

In principle, the maximum number of cases in the meta-analysis would be the number of traits times the number of methods times the number of groups on which the analysis was performed; however, in actuality one does not need to use complete MTMM data designs (if there are more than the minimum of traits and methods, some "holes" are permissible), and one does not have to use all or the same groups for every MTMM data set. In my study the actual number of cases for the meta-analysis was 2,115. (Across my six MTMM data sets, there were an average of about 6 traits and about 3 methods, and estimates were computed for an average of about 20 groups.)

The smallest identified MTMM model depends, of course, on the constraints one imposes. With the constraints I used (method factors independent from each other[1] and from the trait factors, and all method effects from a single factor constrained to be equal), models for four traits and two methods (or two traits and four methods) and models with three traits and three methods are identified. Models with only three traits and two methods may also be identified under these conditions, but I have not rigorously examined this or tried to use models this small. If we take the 3x3 model as an example of a useful small model, it will generate measurement quality estimates for 9 measures, i.e. 9 validity estimates, 9 method effect estimates, and 9 estimates of random error variance.

For any significant meta-analysis, one needs a large number of cases, certainly more than nine. There are three ways to increase the number of cases, and I have used all three in my own work.

An obvious first approach is to combine results from several MTMM data sets. One can combine results from different surveys and/or incorporate more than just one MTMM data design in a single survey.

A second way to generate more cases is to use bigger MTMM designs. The number of traits represented in an MTMM model can easily be expanded in a survey setting. However, survey respondents tend to find repeated questions about the same topic that differ only with respect to the measurement method tedious, so as a practical matter the maximum number of different methods is three or four. Six traits each assessed by four methods would provide twenty-four sets of measurement quality estimates.

A third way to increase the number of cases is to generate measurement quality estimates not only for the sample as a whole, but also for subgroups characterized on the basis of age, and again for three subgroups based on education would provide 63 sets of measurement quality estimates (=3*3*[1+3+3]). In addition to increasing the number of cases available for the meta-analysis, this approach makes it possible to examine the relations between the selected subgrouping characteristics and the measurement quality

[1] Although the research reported in Andrews (1984) consistently constrained method factors to be independent of one another, more recent work I have done with Herzog and Rogers has relaxed this constraint. Allowing method factors to be related to each other has conceptual appeal and sometimes improves the fit of a model. However, when all the traits are also substantially interrelated, an ambiguity arises as to what portion of the relationships among the observed measures should be attributed to correlated traits versus correlated methods. If methods are allowed to relate to one another, it seems important to include at least one trait that is independent of other traits and/or one method that is independent of other methods.

estimates, a topic of considerable interest in its own right. (One could find out, for example, wether more educated respondents provided more precise answers.) Of course, analyzing overlapping subgroups produces some theoretically messy dependencies, but from a practical standpoint the messiness has not (so far) proven problematic. Some new results on this are presented below.

ON DESIGNING THE DATA FOR A META-ANALYSIS AND CHOOSING AN ANALYSIS TECHNIQUE

I have called the analysis of the cases described above a "meta-analysis" or a "stage 2 analysis" because it involves the analysis of measurement quality estimates obtained from prior causal modeling analyses. The meta-analysis is a straightforward analysis that attempts to explain each dependent variable (the estimates of validity, method effect and random error) on the basis of a set of multiple predictor variables (the survey design/content features and/or population subgroup).

As in any such analysis, several aspects of the data need to be considered and will affect one's choice of analysis technique. These aspects have to do with assumptions about (a) multicollinearity, (b) additivity, (c) linearity, and (d) metric variables. The analysis technique that I used, Multiple Classification Analysis MCA (Andrews et al., 1973), was chosen because it can appropriately handle data with substantial correlations among the predictors (multicollinearities), certain nonadditivities, nonlinearities, and nonmetric predictor variables.[2] An alternative but less convenient technique would have been to use dummy variable multiple regression.

Multicollinearity

Multicollinearity refers to the fact that there may be correlations among the predictor variables. Unless the correlations are very high, multivariate techniques such as multiple regression and multiple classification analysis handle it routinely and provide parameters (regression's b's and β's and MCA's effect coefficients) that "hold constant" the effects of all other predictors.

In a study such as this, however, high or even perfect multicollinearity can easily occur and must be guarded against. The

[2] MCA is included in OSIRIS, MICROSIRIS, and SPSS.

problem can be easily illustrated: if the combinations of item characteristics (which are the predictor variables in the meta-analysis) are always as given below, one could not distinguish the effect of the number of categories and the position of the item in the interview because these two characteristics always go together:
- 3 categories and near the beginning of the interview
- 4 categories and in the middle of the interview
- 5 categories and near the end of the interview.

This is an extreme example of overlap that completely confounds number of categories and position in questionnaire. However even the following involves a problematic instance of multicollinearity:
- 3 categories and near the beginning of the interview
- 4 categories and at the middle or end of the interview
- 5 categories and at the middle or end of the interview.

In this case, the effects of using a four- or five-category scale and the effects of being in the middle or at the end of the interview can be distinguished, but the effects of using a 3-category scale cannot be distinguished from a position near the beginning of the interview.

It is helpful to consider this situation as a bivariate frequency table. In table 1, the X's indicate cells in which some cases (survey measures) fall, and the O's indicate cells that have no cases.

table 1: A problematic instance of multicollinearity

	3 categories	4 categories	5 categories
beginning	X	O	O
middle	O	X	X
end	O	X	X

The problem arises because the only cases in the top row also fall only in the left column. Whenever there is a unique row-column combination, the respective categories will be confounded, and it will be impossible to distinguish their separate effects. (Other problems of multicollinearity involving three or more predictors can also arise, but if one has guarded against the occurrence of two-variable problems such as that illustrated above, the more complex ones will be quite rare.)

The "solution" to the multicollinearity problem is, ideally, to prevent it from occurring. This requires that one has control over the design of the data set and that one thinks through in advance the nature of the meta-analysis that will eventually be performed. The literature on experimental designs includes potentially useful suggestions on this matter.

If prevention is not feasible, then one must examine the data one has actually obtained to identify instances of overlapped categories. One can then collapse predictor variable categories or eliminate whole predictor variables to eliminate any problems that appear.

For the meta-analysis reported in Andrews (1984), several predictor variables had to be eliminated and several categories collapsed (as described in footnote 18 of that article). Although all the most serious problems were solved, I have since come to suspect that the surprising empirical results for the "20+" category of number of scale categories might be a multicollinearity artifact involving the predictor that records category labeling.

Nonadditivity

A technique that would handle anticipated statistical interactions (nonadditivities) was needed because it was expected that the effect of the length of the question on validity would depend on the length of the introduction to that question. If a long introduction was used, it was expected that question length might relate negatively to validity, but if the introduction was short the length of the question could relate positively to validity. The underlying idea is a simple one: that there is an optimal amount of information to be presented to a respondent, and that long introductions combined with long questions may present too much information, while short introductions combined with short questions may not present enough.

To allow this interaction effect to be handled within MCA, the two original predictor variables (question length and Introduction length) were combined into a single "pattern variable", and the pattern variable was then used as one of the predictor variables in the MCA analysis. The pattern variable was constructed by first bracketing each original variable into three categories (long, medium, and short, the exact definitions for these appear on page 431 of Andrews, 1984), and then representing the nine possible combinations of these two sets of three categories as a single 9-category variable. Using the pattern variable, the MCA analysis did show the expected interaction.

Nonlinearity

Some nonlinear effects were also expected. For example, the very first questions in a survey are frequently suspected of providing less good measurements than later questions because the respondent and interviewer are just beginning to learn how to work together and establishing rapport. Similarly, the final questions in a long interview may provide poorer measurements because the respondent and/or the interviewer may have become fatigued. Thus one would expect a curvilinear (an inverted-U) relationship between validity and the position of the question in the interview. This did appear in the MCA analysis.

Nonmetric variables

Another problem in the meta-analysis is that many of the predictor variables are nonmetric variables. Some are pure nominal variables (e.g., wether the data were collected using face-to-face interviews, telephone interviews, or group-administered questionnaires) and other predictors had a limited set of ordered categories. MCA is designed for such predictor variables.

SUBGROUP ANALYSIS

In the next part of this presentation I will discuss the subgroup analysis that was included in the original study and present some results that have never been published so far.

As noted above, developing measurement quality estimates for contrasting (and perhaps overlapping) subgroups of the respondents has two attractive features: it provides more cases for use in the meta-analysis and it provides an opportunity to see how subgroup membership relates to the measurement quality estimates.

All together I was able to look at 52 different subgroups. (Adding in estimates obtained from the total set of respondents brings the total to 53 groups.) These are shown in table 2. As one can see, groups were defined on the basis of a wide variety of sociodemographic and survey-processing variables. An attempt was made to define the same subgroups among respondents to each of the six surveys that I used, but this was not always possible and hence some subgroups are defined in only a subset of the surveys.

table 2: Effects of respondent characteristics on data quality

	number of estimates	validity	method effect	residual error
group:				
all respondents together	106	.01[1]	.00	.00
education:				
0-11 years	82	-.04	.01	.06
high school (or HS plus tech.)	63	.00	.01	-.01
some college	63	.03	-.04	-.02
bachelors degree or more	87	.03	.00	-.05
grade school to some college	24	-.00	.01	.00
some college or more	19	.04	-.03	-.03
age:				
18-34 (or 18-30)	106	.02	-.02	-.02
35-54 (or 31-56)	87	.00	-.01	-.00
55-90	82	-.04	.03	.05
65-70	19	-.06	.05	.05
71-90	19	-.08	.11	.03
race:				
white	82	.01	-.01	-.01
black	82	-.04	.01	.04
sex:				
female	106	-.00	.00	.00
male	106	.01	.00	-.01
where respondents grew up:				
rural	24	-.01	-.01	.01
suburban	24	-.01	-.04	.02
urban	24	.02	.01	-.04
seniority in firm X:				
0-4 years	24	.00	.01	-.01
5 or more years	24	-.00	-.01	.00
interviewer's ratings:				
resp.'s interest high	82	.02	-.01	-.03
resp.'s interest low	82	-.03	.02	.02
resp.'s intelligence high	19	.03	-.02	-.03
resp.'s intelligence low	19	-.00	.01	.01
resp.'s sincerity high	19	.02	-.02	-.01
resp.'s sincerity low	19	-.01	.03	.01
resp.'s suspiciousness high	19	.01	-.01	-.02
resp.'s suspiciousness low	19	.01	-.04	-.00
resp.'s reluctance high	9	.01	-.08	.02
resp.'s reluctance low	9	-.00	.03	-.01

table 2: (continued)

	number of estimates	validity	method effect	residual error
respondents own ratings:				
interview seems long	9	.-03	.04	.03
interview seems short	9	.01	-.02	-.01
interest in survey high	9	-.02	-.06	.03
interest in survey low	9	.01	.04	-.03
importance of survey topics high	24	-.00	.02	.00
importance of survey topics low	24	-.00	-.02	-.00
survey's expected impact high	24	-.00	.01	-.00
survey's expected impact low	24	-.00	-.01	.01
assistance by interviewer:				
none or once	12	.02	-.01	-.02
twice or more	12	-.05	.08	.05
clarifications requested:				
none or once	9	.01	.03	-.02
two or more	9	-.03	-.12	.06
questions repeated:				
none	9	-.01	.04	-.00
a few to many	9	.00	-.01	-.01
was resp. interviewed by SRC before?				
no	54	.01	.00	-.01
yes within 6 months	54	.01	-.00	-.00
number of attempts to reach resp.:				
one	73	.01	-.01	-.01
five or more	73	.01	-.01	-.01
special interviewing techniques:				
none, standard methods	9	.01	-.03	.01
spec. instructions, commitment, etc.	9	-.01	.04	-.01
resp.'s concern for social desirability:				
high	51	-.01	.01	.02
low	51	.02	.01	-.03
explanatory power of 53 groups above:				
eta^2 adj.		.12	.16.	.05

statistical significance:
By conventional tests of significance, a difference between these means is significant at the p= .05 level, if the difference is at least .02 and N's are at least 50, or the difference is at least .03 and N's are at least 25, or the difference is at least .05 and the N's are about 10. Standard errors for the coefficients are about .007 when N is 100, .010 when N is 50, and .020 when N is 10; see text.

[1] The coefficients show deviation from the mean associated with membership in the designated category after effects of 13 survey design characteristics have been removed.

Because the subgroups that could be defined differed from survey to survey, and because the design characteristics of the survey measures also differed from survey to survey, it was possible that the group-defining variables might be related to one or more of the design characteristics. This problem was solved by using a two-step analysis: The design characteristics were used first to explain all the variance they could in the measurement quality estimates, then the group characteristics were used to explain as much of the remaining (i.e., residual) variance as they could. The survey design variables explained 66% - 72% of the variance in the measurement quality estimates (as reported in table 5 of my 1984 article), and the group characteristics explained an additional 5% - 16% of the remaining variance (as shown in table 2).

On the whole, the subgroup analysis shows that the effects of the grouping variables are relatively small, but interesting results in expected directions are found in many instances. The largest effects are found for education and age. As one can see, less educated respondents and older respondents tended to provide somewhat less valid data. (Although table 2 does not report a multivariate analysis, another analysis, not shown, indicated that each of the above effects persisted even when the other was held constant.) The results of this analysis stimulated further research into the effects of age. Using other data, results reported by Andrews and Herzog (1986) and by Rodgers, Herzog, and Andrews (1988) show that this "age effect" is replicable, and we are currently trying to identify the conditions that affect its impact.)

One of the concerns that, very properly, has been raised about combining measurement quality estimates from overlapping subgroups analyses in single meta-analysis is that the measurement quality estimates are not independent of one another. This makes it very difficult to obtain a reasonable standard error for the subgroup differences and hence difficult to assess the statistical significance of any effects that emerge. Given the exploratory nature of my own work, I chose to disregard this problem and focus on the trends that emerged. However, recently I have obtained an answer to the question whether the trends shown in the entire data set I used (which included all the overlapping subgroups) would be the same as the trends in a "pure" data set consisting only of estimates from the total groups (i.e., excluding estimates from all the subgroups).[3]

[3] I am indebted to participants at the Amsterdam conference in February 1989 and to my colleague Willard Rogers for pushing this issue, and to Rodgers for performing the analysis.

The answer is a reassuring "yes": the trends shown in a meta-analysis based only on the much smaller but "purer" set of cases from the total groups are virtually identical to those reported in table 6 of my 1984 article. Table 3 presents both the originally published results obtained from the large "messy" data set, and the new results obtained from the small but "pure" data set.

SUMMARY AND CONCLUDING COMMENTS

An abstract discussion about meta-analysis of MTMM data can make the matter sound complex and difficult. However, being clear about a few key ideas and applying standard procedures for the analysis of a dependent variable with multiple predictor variables will resolve most of the problems. This presentation has tried to emphasis these points.

One needs to recall that a "case" for the meta analysis consists of a survey measure as applied to a set of respondents. The measure is defined by a survey question that assesses a particular topic, which we call a "trait" in the multitrait multimethod parlance, and that uses a particular measurement method. Each case is described by a set of variables. The dependent variables are estimates of the measurement quality with which the given trait was assessed using the given method in the given group of respondents. The independent variables include characteristics of the survey design, of the topic being assessed, and of the respondent group.

The analysis itself is straightforward so long as one uses a multivariate technique that is appropriate for the data. There are likely to be multicollinearities among the predictor variables (perfect overlaps must be avoided), nonadditive effects, curvilinear relationships, and categorical predictor variables. Multiple Classification Analysis or Dummy Variable Multiple Regression, each in combination with pattern variables to handle the interaction effects, are appropriate techniques.

Analyzing overlapping subgroups of respondents is a useful way to generate more cases for the meta-analysis and to be able to explore the effects of subgroup membership on data quality. The overlaps among the subgroups create a statistically messy data set, but (so far) no evidence has emerged that the basic trends become seriously biased.

table 3: Comparison of MCA coefficients and MCA β's from meta-analyses using (1) "total" groups only (N=106) and (2) "all" groups (including overlapped subgroups (N=2115)

	concepts		methods		theta (sqrt)	
characteristic	total	all	total	all	total	all
RESPONSE SCALE						
number of categories (β)	*.61*	*.56*	*.88*	*.68*	*.77*	*.74*
2		-.06		.11	.11	.04
3		-.13	-.04	-.05	.23	.22
4-5	.03	.04		.01	-.05	-.06
7	.01	.00	-.03	-.02	.01	.04
9-19	-.00	.01	.19	.21	-.09	-.07
20+ or actual	.13	.14	-.14	-.13	-.21	-.28
explicit DK (β)	*.32*	*.31*	*.56*	*.45*	*.22*	*.30*
no	-.03	-.04	.05	.04	.04	.06
yes	.08	.09	-.10	-.11	-.10	-.14
category labeling (β)	*.32*	*.27*	*.30*	*.28*	*.14*	*.15*
all labeled	-.03	-.04	.03	.04		.03
some labeled		.08	-.08	-.09	-.08	-.10
explicit midpoint	*.03*	*.01*	*.06*	*.06*	*.01*	*.00*
no	.03	.02	.04	.05	-.02	.01
yes		-.01	-.01	-.02	.01	-.01
ITEM CHARACTERISTICS						
absol. vs. comp. (β)	*.30*	*.28*	*.08*	*.15*	*.38*	*.33*
absolute		-.04	.02	.01		.07
comparative	.06	.07	-.03	-.02	-.10	-.11
length intro/quest (β)	*.10*	*.13*	*.37*	*.35*	*.09*	*.10*
sh/sh		-.05		-.01		-.02
sh/med		-.05	.04	.03	.07	.08
sh/long	.00	-.01	-.02	-.03	.02	.05
med/sh	.01	.03	.07	-.08	-.03	-.04
med/med	.04	.06	-.07	-.08	-.06	-.07
med/long	.04	.06	-.08	-.09	.-02	-.05
long/sh	.00	-.00		.05	-.01	-.00
long/med		.00	.06	.07		-.00
long/long		-.03	.06	.07	.05	.06
QUESTIONNAIRE DESIGN						
battery length (β)	*.17*	*.17*	*.24*	*.19*	*.28*	*.44*
1 item (not battery)		.03		-.04	-.06	-.10
2-4 items		.05	.01	.02	-.09	-.11
5-9	-.02	-.03	-.02	-.03	.06	.09
10 or more	-.06	-.07	.09	.10	.08	.14

table 3: (continued)

	concepts		methods		theta (sqrt)	
characteristic	total	all	total	all	total	all
position in qnaire (β)	*.12*	*.13*	*.24*	*.18*	*.11*	*.16*
1-5	-.02	-.01				.03
6-25		-.02		-.02	.04	.05
26-35	.03	.04		-.01	.05	-.07
36-39	.01	.02		-.03	.01	-.03
40-100	.03	.04		-.05	-.06	
101-200	-.03	-.04		.07	.05	.08
201-348	-.04	-.06	.05	.06	.04	.08
data coll. procedure	*.07*	*.03*	*.29*	*.24*	*.01*	*.02*
telephone		-.01	-.00	-.02		.01
F-T-F		.00	.09	.10	.00	-.01
group adm	.00	.04	-.07	-.05	-.02	.04
TOPIC CHARACTERISTIC						
soc. des. sens. (β)	*.07*	*.07*	*.01*	*.00*	*.08*	*.08*
low or medium	-.01	-.02		.00	.02	.03
high	.04	.05	-.01	-.00	-.07	-.08
content specificity	*.08*	*.06*	*.01*	*.00*	*.06*	*.04*
low		-.03		.01	.03	.04
medium		-.01		-.00		.02
high	.03	.02		-.01	-.04	-.03
experience vs. predict	*.01*	*.01*	*.00*	*.00*	*.01*	*.01*
actual experience		.00	.00	-.00		-.01
predictions		-.02		-.00		.04
content salience	*.01*	*.01*	*.00*	*.00*	*.00*	*.00*
low	.00	.01		-.01		-.01
medium		.01		-.00	-.00	.00
high		-.01		.01	.00	.01
R^2	.69	.66	.88	.72	.66	.67

A blank in the "total" column means MCA coefficient matches the "all" column.

Meta-analysis of measurement quality estimates can provide extremely interesting answers to the questions of to what extent (and in what ways) does survey measurement quality relate to the design and administration of survey research.

REFERENCES

Andrews, F.M. (1984). Construct validity and error components of survey measures: A structural modeling approach. *Public Opinion Quarterly, 48,* 409-442.

Andrews, F.M., & Herzog, A.R. (1986). The quality of survey data as related to age of respondent. *Journal of the American Statistical Association, 81,* 403-410.

Andrews, F.M., Morgan, J.N., Sonquist, J.A., & Klem, L. (1973). *Multiple classification analysis* (2nd ed.). Ann Arbor, MI: Institute for Social Research.

Rodgers, W.L., Herzog, A.R., & Andrews, F.M. (1988). Interviewing older adults: Validity of self-reports of satisfaction. *Psychology and Aging, 3,* 264-272.

chapter 10

The experimental approach for the meta-analysis

G. MELLENBERGH

In his contribution, Andrews has presented a typical non-experimental approach for the meta-analysis. In this chapter I would like to suggest the use of an experimental design. The advantages are that one knows in advance which effects are relevant and can be discussed and which are not. In fact, we can plan a study in such a way that we can determine in advance which factors are relevant.

In this contribution I do not want to go into details but will only give a rough idea of the possibilities.

BASIC CONCEPTS

In order to discuss the topic some basic concepts have to be introduced. First of all, we speak about "factors" as the variables which are seen to be independent and the scores on these variables are called "levels".

The terms "complete crossing" and "partial crossing" are used in order to indicate whether the levels of one factor are combined with all levels of the other factors or not. For example, the design problems mentioned earlier by Andrews are typical examples of partial crossings.

Furthermore, a distinction is made between "experimental factors" and "non-experimental factors". The experimental factors can be varied by the researcher whereas the non-experimental factors are not under his control. In the project discussed here, typical experimental variables are concepts and methods, while country and language are non-experimental factors.

Finally, a distinction should be made between variation *between* and *within* respondents. Certain factors can be varied for each respondent: examples are the topic and the method. Such factors produce *within* respondent variation. Other factors, such as country and the data-collection method, cannot be varied for each respondent as a respondent lives in one country and, in general, it will not be possible to ask him the same questions using different data-collection techniques. The last two factors will therefore produce *between* respondents variation.

A POSSIBLE DESIGN FOR THE META-ANALYSIS

There are of course many different possible designs which could be used. Furthermore, the number of variables in this study is very large and it is not possible to work out the final design of this study in such a short time. What I would like to do is to give an example of a possible approach to developing such a design and to give an idea of the possibilities which such a design provides with respect to the estimation of the different effects.

I have chosen a "Balanced Incomplete Block Design" as a reasonable possibility. This design is especially useful where there are a large number of experimental variables as in this research project because most relevant effects can be obtained without a complete crossing of all factors. A complete crossing of all factors would be impossible in this specific case. For other possible designs, see for example Cochran and Cox, 1957; Cox, 1958; Edwards, 1968; Kirk, 1968; Winer, 1962; Collier and Hummel, 1977.

Examples of factors which produce between respondents variation are the countries and the data-collection method. Within variance is introduced by only three different design factors in order to keep it simple. For illustrative purposes I have chosen the length of the introduction to the question, the length of the question itself and the type of response scale used. The factor "countries" has 14 levels, representing the number of countries participating in this research project. The factor "data-collection method" could, for example, have two levels: face-to-face and telephone interviewing. Consequently 28 conditions can be defined by a complete crossing of these two factors. The within respondent factors each have two levels. For the length variables the levels are short and long. For the type of scale we could also use two levels, for example, category scales versus continuous scales. On the basis of this 2 by 2 by 2 design, 8 different combinations of within respondent factors can be distinguished.

Using this design and specifying that in each study at least 4 different design methods are used, one can formulate the experimental design matrix presented in table 1.

table 1: Balanced incomplete block design with four methods per study

length of intro:		short				long				
length of question:		short		long		short		long		
type of scale:		ca	co	ca	co	ca	co	ca	co	total
		1	2	3	4	5	6	7	8	
country	data coll.									
1	1	x	x	x	x					4
	2	x	x	x	x					4
2	1					x	x	x	x	4
	2					x	x	x	x	4
3	1	x	x					x	x	4
	2	x	x					x	x	4
4	1			x	x	x	x			4
	2			x	x	x	x			4
5	1	x		x			x		x	4
	2	x		x			x		x	4
6	1		x		x	x		x		4
	2		x		x	x		x		4
7	1	x			x		x	x		4
	2	x			x		x	x		4
8	1		x	x		x			x	4
	2		x	x		x			x	4
9	1	x	x			x	x			4
	2	x	x			x	x			4
10	1			x	x			x	x	4
	2			x	x			x	x	4
11	1	x		x		x		x		4
	2	x		x		x		x		4
12	1		x		x		x		x	4
	2		x		x		x		x	4
13	1	x			x	x			x	4
	2	x			x	x			x	4
14	1		x	x			x	x		4
	2		x	x			x	x		4
total		14	14	14	14	14	14	14	14	112

It is typical of this design that the number of methods in each row is equal for each row. Each method (column) also appears with equal frequency. Even all combinations of methods occur with equal frequency. However, not all possible combinations of methods have been used in each country as this would have meant that in each country 70 combinations of 4 methods out of 8 possibilities would have to be tried for each method which is of course impossible. In this design only two combinations are used in each country but the design allows conclusions about 8 different combinations.

With this design in each country one experiment with personal interviewing would have to be done and one by telephone and in each study 4 different methods would have to be used. If this design across countries was chosen, where each country does two studies with specific combinations of methods, then conclusions can be drawn for all possible combinations of methods. How these conclusions can be derived will be explained below.

META-ANALYSIS

The dependent variable in this study is an estimate of a validity or invalidity coefficient, or an error variance. In all cases the original variable or the square of the statistic is thus an R^2. Therefore, it might be better to apply Fishers z transformation or a logit transformation on the statistic and use the transformed scores as the dependent variable.

For each cell of table 1, a value of this dependent variable can be obtained. Analysis of variance can be used to analyze the scores. This will lead to the following result:

$$z(R^2) = \mu + \alpha_i + \beta_j + \alpha\beta_{ij} + \gamma_k$$

This allows an estimate of:

- the mean value (μ)
- the effect of the country (α_i)
- the effect of the data collection method (β_j)
- the interaction between the two ($\alpha\beta_{ij}$)
- the effect of the different methods (γ_k)

CONCLUSIONS

In this contribution, I have discussed the possibilities of experimental designs for this research project. This approach has been taken because the project still has to be designed. Therefore the data collection can be planned for more efficiency than a non-experimental design would allow.

The advantages of such an approach are that it will not be determined after the data collection is completed that very little can be said about the effect of certain factors. In addition the different possible designs can suggest economic ways for reducing the number of studies which have to be done in order to estimate the effects.

The trade-off is that one has to spend a lot of time in designing the study. The organization of the data collection is also more detailed than in the non-experimental approach.

REFERENCES

Cochran, W.G, & Cox, G.M. (1957). *Experimental designs*. New York: Wiley.

Collier, R.O, & Hummel, T.J. (1977). *Experimental design and interpretation*. Berkeley: McCutchan Publ.

Cox, D.R. (1958). *Planning of experiments*. New York: Wiley

Edwards, A.L. (1968). *Experimental design in psychological research*. New York: Holt, Rinehart and Winston.

Kirk, R.E. (1968). *Experimental design: Procedures for the behavioural sciences*. Belmont: Brooks/Cole Publ.

Winer, B.J. (1962). *Statistical principles in experimental design*. New York: MacGraw-Hill.

chapter 11

Wording variables for meta-analysis of MTMM studies

N. MOLENAAR

Introduction

The reason why I have been invited to present something relevant for the project under discussion, is - I guess - the fact that some years ago I did research on the possible effects of formal characteristics of survey questions upon the responses given to such questions (Molenaar, 1986). I am not going to discuss here the intricacies of my own research; much of it would be irrelevant to you. Neither will I discuss the results of my research, however tempting that may be. But I have to tell you something about my research, because it will provide the background to the comments I am going to make about aspects of the excellent article by Andrews (Andrews, 1984), in particular about the predictor variables he used in his study.

BACKGROUND OF THE SURVEY QUESTIONS RESEARCH

The aim of my research was to find out which formal characteristics of survey questions (wording aspects or "wording variables" as I like to call them) are the most important in determining the answers respondents give to these questions. My research did not contain any validity measure; i.e., I did not have data from which one could positively derive how survey items should preferably be worded to get the most valid answers.

This aim of mine was not new at all, but the method was my own. The design of my research was non-experimental. I selected 500 survey questions (items) out of the thousands that were stored in the archives, together with their corresponding response distributions, which were also available in the archives. All these 500 questions were attitudinal, closed, posed in face-to-face interviews, and administered to large nationwide samples of respondents, to mention only some of the criteria I used for their selection. Then all

500 question texts were coded as completely as possible, using 12 variables to measure their content, and 25 variables to measure their formulation aspects. The response distributions were also coded: in fact I coded their means, their standard deviations, and the percentages of non-substantive answers (i.e., Don't Know plus No Answer responses). Using a multivariate technique of analysis, I then established the direct relationships between each wording variable (predictor) and each of the three distribution characteristics, holding all other predictor variables constant.

The major point to note is that my research design was non-experimental. In contrast to an experimental design, the exact wording of all my questions was already given; my only task was to code these given texts, or, if you prefer, to "measure" or to "score" them on a set of variables. But, which variables? Which codes? There are so many aspects to question texts, and each question seems to be unique in one respect or another.

Two main problems were encountered in searching for the appropriate wording variables. Both seem to me of general interest for those who plan systematic investigations of the effects of formal question properties, whatever the research design employed.

TWO GENERAL PROBLEMS IN METHODOLOGICAL RESEARCH ON QUESTION WORDING EFFECTS

Search for generic wording variables

In my search for the wording variables I thought that knowledge of the experimental literature would be useful. That, however, was a naive expectation. Most of the variables that were manipulated in those experiments could be recovered in my question texts only after serious reformulation, if at all. Furthermore, there were many more question properties to be distinguished. Therefore, my main strategy for discovering text variables was to compare any two or more questions and to find out in what respects their texts actually differed (apart from their being different in content). Well, it is relatively easy to point exactly to the parts of the text which differ, but not to translate these literal text properties into values on a sufficiently general wording variable. This is the point I would like to emphasize.

To study the effects of question properties systematically, you have to define "generic" variables, i.e., variables which are broad enough to cover several manifest text-properties, but at the same time homogeneous enough to constitute a variable with unambiguous values. The relationship between these two requirements, broadness and precision, or generality and detail, is strained: raising the one lowers the other. The art is to find a proper balance. I attached much value to the precision of the variables for an obvious reason: when opinion researchers apply the results of methodological studies on the wording of questions in formulating their own concrete questions, the wording variables should be as precise and unambiguous as possible. In some cases it was impossible for me to construct such a "generic" wording variable. A small example will illustrate this point.

It is commonly said that one should be very careful in using negations in survey items, and that they should be avoided if possible. Experiments can be designed to investigate whether including a particular negation in a particular place in a particular survey item would have any effect on the responses. There is nothing wrong with such an experiment. The variable that is manipulated may be called "negation in item or not" or perhaps "number of negations in item". All possible survey questions can be coded into this variable. However, the variable would be very heterogeneous, perhaps even "vague". Negations can have different syntactical functions in a text, they can be placed anywhere in a text, and the scope of a negation may be very different. These differences are not covered by the variable "number of negations", which implies that all negations are equal. The variable is broad but not precise. It was, however, impossible for me to construct a "negations-variable" in such a way that it would account for these differences. This example may seem to pay too much attention to small details. Perhaps it does, but this gives rise to the question: how far should one go into details to meet the requirement that the variables should be unambiguous? Later on, when I discuss some of the variables that Andrews used in his study, we will see other examples.

Form and content

The second problem is a general one relating to whether some part of a question text concerns the content of the question or "merely" its wording.

In the negation example would including a negation in an item just change the form of that item or would it change the content? I think, that a change in form would result, but I cannot argue why. The borderline between form and content is not clearly defined. For some aspects of questions it is difficult to decide whether they concern the content or the form. They seem to lie in a twilight zone.

An example will illustrate this point as well as the precision problem just discussed.

Given the following two questions:

1. Do you approve or disapprove of a married woman taking paid employment?
2. Do you approve or disapprove of a married woman taking paid employment while there are small children at home?

I reduced them to the forms:

1. A married woman takes paid employment. Approve/disapprove.
2. A married woman takes paid employment while there are small children at home. Approve/disapprove.

a. Question (2) is longer than question (1). The difference can be covered by the variable "length of the question".
b. Question (2) contains a subordinate clause, question (1) does not. The difference can be covered by the variable "subordinate clauses in item or not".
c. The subordinate clause carries an argument for respondents to disapprove of the statement in the main clause; the first question lacks such an argument. The difference can be covered by the variable "use of counter arguments or not".
d. Many other subordinate clauses are possible: "... when the children are at high-school", "... when it is financially necessary", etc. These two subordinate clauses present an argument for respondents to approve of the statement in the main clause. The variable can be constructed as "use of arguments for or against versus no arguments".
e. The arguments in the subordinate clause may also vary in their strength. It may be possible to "measure" the strength of the arguments in either direction and to construct the variable "strength of the arguments" in either direction and to construct the variable "strength of positive or negative arguments in item versus no argument" or something of the kind.

In all these analyses, the difference between the two questions is considered to be formal. But it can be argued that the difference between the two questions is in the content, or that the questions are not two different versions of one and the same question, but two different questions! It is interesting to note that there is no sufficiently general and usable definition of what formal properties of questions are, in contrast to content. The class of wording variables is not clearly defined; we have to rely heavily on our own intuitive (and subjective) understanding of language.

I would now like to make some comments on the predictors Andrews used in his study. It is not my intention to criticize his work; his article is a useful starting point for my comments about possible wording variables. Most of my comments are essentially guided by the idea I have just discussed: predictor variables should be sufficiently general to cover a variety of concrete (manifest, unique, literal) wording aspects, but at the same time they should be precise.

COMMENTS ON PREDICTORS USED BY ANDREWS

Number of scale categories

In the research of Andrews the variable "number of scale categories" has the largest effect: as the number increases, so does the data quality. I also included this variable in my research. But what does this variable really means? It appears to be clear-cut, is simple and reliable to measure: you just have to count the number of response options. In practice, however, the variable is not so simple: varying the number of answer categories is not just simply lowering or raising an integer; it often entails changing one or more of the basic properties of the scale, especially when the categories are verbally labeled. What matters is which categories have been omitted or added. The problem is quite similar to that in the example about the "negations". In other words, you can construct questions that all have the same number of categories but are nevertheless different.

1. In a checklist question (nominal level of measurement) the number of categories may, for example, be 10. When you leave out three of them, for whatever reason, the number decreases to 7. But, this is not the crucial point. It matters which three categories you leave out. You may leave out check items that are important in that they would have been frequently chosen by respondents, or you may leave out unimportant ones. The number of categories has been changed, but the real variable that has been manipulated is the scale property "completeness". Categories are different in their importance, their saliency, and their familiarity to respondents.

2. Ordinal level scales are in a sense more interesting (they are used by Andrews). One can, for example, raise the number of answer categories from two ("good/poor") to five ("very good, good, neutral, poor, very poor"). It is obvious that the respective scales

differ in their number of categories. But that is not all. If there is any effect due to changing the number of categories, it would be at least partly attributable to the fact that in the five-option format one of the categories happens to be a neutral category. In fact, the response scale has been changed in two respects at the same time: the number is raised and a middle category is added.

3. Questions may have the same number of categories, but be presented in a different sequence (from positive to negative or vice versa). Is this just a detail? For me, the answer is no. Reversing the sequence may - under some conditions - have an effect on the answers obtained (primacy or recency effect). In my research I coded the number of answer categories as well as their sequence. I noticed that there seems to be a widely held convention to present the categories from the most positive to the most negative, and in a linear order.

4. It is easy to imagine that changing the number of answer categories may make symmetrical scales negatively "skewed" (or the reverse); equally spaced scales may become more "stretched" or more "compressed". This can occur in several degrees and ways, depending on how many and which categories are added or omitted. The most common situation is that categories are symmetrically added to or removed from the ends of the scale, making the "width" of the range of the response scale greater or smaller (or the fineness or gradedness). One can also change these scale properties without changing the number of categories. The variable "number of categories" is therefore a rather complex or heterogeneous variable and the implied scale properties are difficult to define in proper variables.

In my own research I included the number of answer categories as a predictor variable, and their linear sequence. I also measured the extent to which a set of categories was skewed (i.e., out of balance or asymmetrical) in the positive or negative direction with respect to an assumed midpoint.

Inclusion of the DK-option (NO-option)

This variable was the second most important in Andrews' research.

1. The DK-option should be included as it raises the validity of the survey measures. Conversely, it will also raise the number of DK-responses, which is generally equivalent to raising the number of "missing cases" in the analysis. The strength of the effect of including the DK-option, in particular the effect on the percentage of DK-responses (but possibly also on the distribution of the sub-

stantive responses) will depend, in my opinion, on whether the interviewer is instructed to mention the DK-option explicitly or not; i.e., whether or not the interviewer calls attention to the DK-option as a legitimate response.

2. The DK-option is a device for filtering out respondents who do not have a substantive opinion about the subject matter. It is one of a range of possible DK-filters and is in fact the weakest. There are, at least in theory (in practice they never occurred in my research) stronger DK-filters (Bishop et al., 1983; Schuman and Presser, 1981). For example: "Not everyone has opinions on these questions. If you do not have an opinion, just say so". Using stronger DK-filters may raise the validity of the survey measures, but at the same time they will also raise the number of DK-responses and, consequently, the number of missing cases in the analysis. This is the serious drawback of raising the validity.

3. In most cases, the DK-option is a response option in its own right, with an apart status. But it is also possible to insert a "No Opinion" category into a response scale (meaning: "Can't decide") as the neutral midpoint of that scale. The placement of an option apart or within a scale is not just a detail.

Battery length

In Andrews' research this variable measures whether the target item was included in a battery with other items and, if so, how long that battery is. The longer the battery, the lower the data quality. It is interesting to note that Andrews did not include the variable that measures how far into a battery the item was located. This last variable may show a weaker relationship to each of the quality components than the former (surprisingly, indeed), but in my opinion it is more comprehensible.

In my research I did not measure battery length. I measured the sequential number of the target question within a series of preceding questions that all had identical answer categories, whether these questions were recognizable as members of a battery or not. This variable is very similar to the variable omitted by Andrews. The higher the sequence number, the higher the number of DK-responses tended to be.

I included this variable (sequential number) in my research because it was easy to measure. But I realized from the beginning that there was something more. I would have been able to code the content of the previous questions, but I was not able to do so without making the research extremely complex. Apart from this technicality where should one start measuring the context of a target question? Would it suffice to measure the content of the question immediately preceding it? Or of the two preceding questions? In some experimental research, even very remote questions (or rather the respondents' answers to them) could be relevant for respondents when they answered the target question (Bishop, 1987). The concept "context of a question" is very complex to define and measure. It matters, for example, whether the previous question is on the same or at least a related subject matter (in the perception of the respondents). Furthermore, there are several different types of context effects: saliency effects, reference group effects, consistency effects, contrast effects, rapport effects, and whole-part effects (Molenaar, 1982). All these possible effects require careful coding of previous questions on different aspects.

Absolute versus comparative

The comparative type of question is favourable to the type of question that calls for an absolute judgement. In the first version of my research I included the variable "comparative - absolute" (which did not have any substantial effects). But in the second version I removed the comparative questions altogether. There were two main reasons for doing so (remember my non-experimental design). The basic form of a comparative question is "Is A better than B"?

The basis of the comparison, B, (the "anchor-point" as Andrews calls it), may in practice be very variable in content. Also, B is assumed to be the basis of comparison for giving an opinion about A. But it is easy to construct another comparative question by switching A and B.

All these questions are of the comparative type, but they are undoubtedly quite different. They may all have their own effects, and possibly the content of the questions will be different. In fact, the term "comparative" is ambiguous. I was unable to construct a variable (or a set of variables) that would cover the possible variants of comparative questions.

Length of the introduction and the question text

Andrews splits the length of an item into that of the introductory and that of the question text. In my research it was sometimes difficult to assess where the introductions and questions began and ended; sometimes they were woven together in one long sentence. But this is only a minor point. I measured the length of the introductions too, but not the length of questions. My comments therefore will concern only the introductory texts.

The length of an introduction (in number of words) is a very broad variable. What matters is the kind of information given in the introduction. The major distinction made in the literature is between non-substantive (fillers, clichés; Laurent (1972): raising validity, higher reportings) and substantive information (possibly leading to misunderstandings; see Belson, 1981). However, I was unable to measure whether the information in the introductory texts was substantive or not. Instead, I tried to code this information (if any) in a set of variables.

1. Some parts of the introduction are incentives to stimulate the respondent to answer the following question, whatever the answer may be. Such incentives can be formulated in several ways, differing in their urge for an answer. They can be formulated as a polite or business-like request. For example, "I will read you some questions now. Would you indicate if you agree or disagree?"; "I will read you some questions now and each time you say if you agree or not".

The incentives may also be formulated as a command or as a cautious, friendly request. For example, "I will read you some questions and you must say if you agree or not"; "Would you be so kind to tell me if you agree or disagree?"

2. Other parts in an introductory text emphasize that a respondents' subjective opinion is asked for in the question. (A weak device for personalizing questions). I coded the number of times such words or parts of a sentence occurred. Examples are: "Do you think ...", "... in your opinion ...", "according to you", "... your own opinion". In some questions these occurred up to four times.

3. Sometimes response categories are mentioned in the introductory text as well as presented with the questions. If present, all of them may have been mentioned, but often only a subset is presented. I coded these extra response categories in exactly the same way as I coded the normal response categories (which are commonly printed somewhat apart from the question in the questionnaire). I measured the extent to which the set of extra response categories was skewed in the positive or negative direction. For example "I would like to know if you agree with the following statement (... statement ... ; response categories)". The word "agree" in the introduction is "extra", it is also a skewed extra (only the agree option is mentioned). Whether the skewness is in the positive or negative direction will depend on particular features of the question (or statement) itself.

4. Introductory texts may also contain constructions such as "Some people think ... other people think ...". I measured the degree

to which the opinions of these "other people" in such constructions are given unequal weight in the positive or negative direction. It is easy to grasp the meaning of this variable, when you compare the following three constructions:

a. "Some people say (... something positive ...). What do you think?"
b. "Most people think (... something positive ...). What do you think?"
c. "Some people think (... something positive ...), other people think (... something negative ...). What do you think?"

In this last construction the positive and the negative side of the possible opinions are given equal weight; the other constructions are skewed.

5. Knowledge that is judged to be relevant to respondents for answering the question may be transmitted. Its function is to inform respondents who might not know what the following question is all about. Sometimes the information is given in a straightforward way, but mostly the knowledge-transmitting sentences start with something like "As you may know, President Reagan has recently ...". Note the "As you may know ...".

6. Most parts of introductions are texts for interviewers to read aloud to respondents when they have to handle showcards. I did not code these parts of the introductions. But I did include the variable "showcard used or not" and, if so, what information was on the card: only the answer categories; only a part of the question; or, the answer categories plus a part of the question. This order reflects the increasing amount of written information respondents have at their disposal when answering the question.

Position of the item in questionnaire

I also included this variable in my research. It roughly indicates the time elapsed since the first question was asked. It was sometimes difficult to establish the position of an item in the questionnaire, namely when there were several subroutines present. There is something curious about this variable: some questions have to be first in a questionnaire, and some have to be last. If the position in the questionnaire is correlated with data quality, it follows that some questions will always produce data of lower quality than other questions.

Category labeling

This variable was not included in my research: all answer categories of my 500 questions were verbally labeled. But I did measure how they were labeled:

a. categories labeled with full, well formed sentences (i.e., complete repetitions of the question text)
b. categories that only repeated the predicates in the question (for example, "important/not important"; "good/bad")
c. categories with only short, abbreviated terms that were not comprehensible without reverting to the question text ("yes/no/sometimes"; "agree/disagree").

Neutral midpoint

I included this variable in my research as well, but in a slightly different way to Andrews (there is of course the problem when a category is to be considered as a neutral middle category): I measured the number of times such a neutral middle category was mentioned in the text.

In fact, many more wording variables were included in my research. I will only mention them briefly:

a. The readability of the question.
b. Whether the question is formulated in an interrogative mode or as a statement (declarative sentence).
c. The type of the question: "yes/no" or "choice".
d. The degree to which the set of predicates that are given in the question text is skewed.
e. The degree to which the formulation of the attitudinal object of the question is positively or negatively biased.
f. The degree to which the question is positively or negatively loaded by the presence of clauses or emotionally-toned words.
g. Whether the question is hypothetical or not.

CONCLUSIONS

In choosing the wording variables to be included in systematic methodological research (I assume that researchers have full control over that decision), one may prefer to choose the question properties that are generally considered to be formal wording variables and/or can be easily (and reliable) measured. You can choose broad but less precise variables or precise but narrower variables.

One should find a proper balance between broadness and precision. It should be borne in mind that in the end the results of methodological research are intended to be applied by researchers in the field when they have to formulate their own questions.

REFERENCES

Andrews, F.M. (1984). Construct validity and error components of survey measures: A structural modeling approach. *Public Opinion Quarterly, 48,* 409-442.

Belson, W.A. (1981). *The design and understanding of survey questions.* Farnborough: Gower.

Bishop, G.F. (1987). Context effects of self-perceptions of interest in government and public affairs. In H-J. Hippler, N. Schwartz and S. Sudman (Eds.), *Social information processing and survey methodology.* New York: Springer Verlag.

Bishop, G.F., Oldendick, R.W., & Tuchfarber, A.J. (1983). Effects of filter wording in public opinion surveys. *Public Opinion Quarterly, 47,* 528-545.

Laurent, A. (1972). Effects of question length on reporting behaviour in the survey interview. *Journal of the American Statistical Association, 67,* 298-305.

Molenaar, N.J. (1982). Response-effects of "formal" characteristics of questions. In W. Dijkstra and J. van der Zouwen (Eds.), *Response behaviour in the survey interview.* London: Academic Press.

Molenaar, N.J. (1986). *Formuleringseffecten in survey-interviews (Wording effects in survey-interviews).* Amsterdam: VU-Uitgeverij.

Schuman, H., & Presser, S. (1981). *Questions and answers in attitude surveys: Experiments on question form, wording and context.* New York: Academic Press.

chapter 12

Summary of the discussion

W.E. SARIS

The two presentations on the design of the meta-analysis have shown that this task should not be taken lightly. In order to estimate the different effects, a great deal of attention must be paid to the design of the study. In this respect experimental designs should be looked at carefully.
On the other hand, a non-experimental design would allow some flexibility, which is also required in such a complex international research project. We cannot specify exactly how many studies should be done, nor in which form they should be carried out by the different participants. This is certainly not possible at this moment when all participants have only just started to look for funding of their research. The subsidizing organizations will obviously also have some effect on the kind of research that will be done.

It is therefore probably more realistic to discuss the necessary experiments while allowing a lot of freedom to the different participants in the choice of their studies. However, we will register the procedures used very carefully and after some time we will study the variation in experiments which have been done so far. Experiments which are necessary to obtain as much information from our data as possible will then be specified.

During the discussions it was agreed to delay the decision of the design of the study for the moment. We shall begin with a pilot study which will be done in all countries in order to get acquainted with the project and to determine the different steps.

Some discussion was held about the merits of MCA over ordinary regression in the analysis. The final conclusion was that MCA would in principle provide the same information as would be obtained for regression analysis with dummy variables. It therefore does not matter which of the techniques we use in our analysis.

The following specific points were also discussed:
- the dependencies which occur by necessity in the design
- a possible division of labour.

THE DEPENDENCIES IN THE DATA

Andrews made it clear that at least two dependencies will arise in this design. First of all, we can expect dependencies between the qualitative measures of the measurement instrument because of the covariation of the estimates of the parameters in an MTMM model. This has been discussed by Boomsma (1984). In some cases these correlations between the estimates can be very high. If restrictions are introduced on the parameters, the relations exist by definition. Such restrictions have been suggested by Andrews for the method effects or invalidity coefficients. He suggested that these effects be constrained so that they are equal. In the meta-analysis this means that some cases will, by constraint, have the same value for the dependent variable. We should consider whether to reduce the number of cases in order to take this dependency into account.

If no equality constraints are made, the dependencies will still exist but they will be less extreme. Information about the strength of the relationships between the estimates is also available in the information matrix which LISREL provides. Some further study is required to determine how these dependencies should be corrected for.

The second dependency which will occur with this design is created by the subgroup analyses. Many different subgroups will have to be analyzed, each time based on partly the same respondents. Due to this overlap of respondents in the different subgroup analyses, dependencies between the cases in the meta-analysis will also appear. The results may even be inflated by the frequent use of the same respondents. The reason for subgroup analysis is clear. The background variables can affect the results and these effects are certainly interesting. Furthermore, the subgroup analysis produces many more cases for the meta-analysis.

The last argument is possibly not so important in our study if we can organize a large number of studies across the different countries participating. The first point still holds. It was suggested that all countries could do such subgroup analyses on their own data and report on the results of these studies without having to include them in the final meta-analysis. In this way the last type of dependency between the cases in the meta-analysis can be avoided. We shall then only have to find a solution for the first type of dependency in our study.

CHAPTER 12

DIVISION OF LABOUR

Data has to be collected for:
- individual respondents
- measurement instruments.

For respondents, different kinds of information have to be obtained:
- personal characteristics of the respondents
- characteristics of the interviewer
- characteristics of the interview process

For measurement instruments, the following information has to be collected:
- characteristics of the question
- characteristics of the response scale
- survey design aspects
- topic of the question

The criteria to be used in the description and the topics to be used have been mentioned above.

These two separate files should be analysed and then combined in one new data file for the meta-analysis over all instruments. In order to arrive at this meta-analysis, an analysis on the individual level file has to be done to derive the quality of the instruments and the characteristics of the sample. This indicates that a lot of work has to be done with respect to data analysis in each country. A lot of work is also necessary with respect to the analysis but it was suggested that it be divided between the participants and the co-ordinating research team in Amsterdam.

The analysis task of each research team

Two kinds of analyses of the individual data files are necessary:
1. an analysis to determine the quality of the measurement instrument
2. an analysis to determine the characteristics of the sample used

In the analysis to determine the quality of the measurement instruments we suggest that the following steps be included:

a. data cleaning on outliers by inspection of scattergrams and response patterns of individual respondents.
b. determination of the sample to be used (possibly sub-samples of the original sample).
c. calculation of the Pearson correlation coefficients for the variables of interest using pairwise deletion.

d. estimation of the quality indicators using ML estimation on the basis of the above mentioned correlation matrix and the model chosen by the research group (MTMM model).

In the analysis to determine the characteristics of the sample used the researcher must do a frequency analysis of the data on these characteristics. A proposal for the categories to be used will be made in a computer assisted questionnaire. The result should be a list of characteristics of the (sub)-sample used in terms of, for example,

- the percentage of respondents with less than 6 years education
- the percentage of respondents with 6 or more years of education but less than 10,
- the percentage of respondents with 10 or more years education.

In the same way, the sample can be described by other characteristics. It is important to describe the sample as used in the analysis of the data and not the original sample or the hypothetical sample without refusals.

Analysis task of the Dutch research group

In order to formalize the procedure and present the different items of information which are necessary for the meta-analysis, each research group will be provided with a diskette (+ backup) containing the interview program INTERV and an interview which must be answered for each study analyzed.

The interview will ask for:

1. the name of the researcher
2. the name of the study
3. a short verbal description of the sample
4. the number of topics in the sample used for the project
5. the characteristics of the sample
6. the number of questions per topic
7. a description for each topic in the characteristics specified above
8. the characteristics of each topic for each question

By completing this interview, all researchers will provide the information in the same format and in machine-readable form. The program produces an ASCII file which can automatically be transformed to a SPSS, BMDP or SAS file by a special program (SYSFILE). The transformed data file will be analyzed to detect the effects of the different factors. This meta-analysis will be done in

Amsterdam. The data should be sent to Amsterdam on a floppy disk containing:

- the original cleaned data and the questionnaire (paper or computer file)
- the correlation matrix used and the obtained output (paper and/or disk)
- the response file of the interview (diskette)

In return, the Amsterdam centre will provide any necessary information or help which is requested by participants.

DISCUSSION OF THE CHOICE OF THE VARIABLES

The contribution of Molenaar had a sobering effect on the conference participants. He made it very clear how difficult it is in practice to study the effects of the kind of methodological factors we are interested in. Billiet suggested that the approach which Molenaar had followed was partially the cause of the problems he encountered. Coding all variations in the formulations of questions is a very difficult task. In an experimental or quasi-experimental design the researcher has more control over the choice of the wording and the scales. The researcher will thus also have more control over the factors which are varied by him. Billiet presented a number of striking examples from his own research where he found an effect due to question and scale wordings on the distributions of the responses. These examples are contained in a recent publication, which is unfortunately only available in Dutch (Billiet et al., 1984).

In our project we shall of course not be studying the effects on the distribution of the responses but on the validity, invalidity and random error. Billiet thought it was likely that many variables which have been found to have an effect in his research, in that of Sudman and Bradburn (1974), Schuman and Presser (1981), and of Molenaar (1986) will also have an effect on the qualitative characteristics we are interested in. He mentioned that he was especially interested in the effect of different scales on qualitative characteristics. He expected that certain response patterns which occur for specific scales will lead to correlations between questions measured with the same kind of scales.

During the conference there was a general feeling that we should concentrate on the variation in scales. Gehmacher thought that the formulation of questions will always remain an art, therefore we should rather concentrate on the part for which there is some kind of apprenticeship, for example the scaling procedures. Roe mentioned that the measurement of social science phenomena was so bad that it was timely to see how this could be improved and he expected a major improvement from the use of continuous response scales. Melich gave an illustrative example of the importance of this research. In the Eurobarometer some questions were recently changed from 4 points to 10 points scales and now the results were quite different. The effects seemed to be different for the different countries as well.

Saris also mentioned an interest in the effect of different scales. He expects that large scale studies will show the better quality of continuous scales that has so far only been found in small scale experiments. He is also interested in the effect of different procedures for using reference points on the scale. He argued that people can express themselves in different ways: some use overstatement and others understatement while they may have the same opinion. Different words could therefore express the same opinion and the same words could express different opinions. He has done many years of small scale research to test these hypotheses and to find ways of correcting data affected by these problems but has been unable to find a proper correction procedure. The only solution he has found is the use of scale reference points which fix the relationship between the subjective scales and the response scales. A commonly used procedure by his group for example was to instruct the respondents in satisfaction studies to use a scale of 0 to 1000 where

1000 = ideal
0 = nothing in common with the ideal.

With this instruction it could be shown that the variation in the response functions could be reduced considerably (Saris, 1988).

Roe mentioned, however, that in his media research the category scale was compared with continuous scales from 0 to 100. For some variables he found more extreme responses than he had ever seen for the categorical scales. He wondered whether this effect might be due to the use of only two reference points at the end points of the scales as was suggested by Saris.

It was clear to the audience that on this aspect of the scaling instructions further research was also necessary. Although the variation of scales is certainly important, we nevertheless have to take into account that the formulations of the questions will also vary and this cannot be ignored in the research design.

There was also considerable interest in the effect of data collection methods even though the experiments might be limited because not all participants can experiment with data collection methods in the way this is possible in the Netherlands. Some division of labour is required in this respect.

A topic of general interest was the frequency question. Schadee argued that this question should be included because it is so widely used for many topics whereas there is little knowledge of the effect of the question formats. Even though we are happy that many of these questions, which are notorious for response errors (Belson, 1981), can be avoided in fields like media research because electronic devices are often better at registering behaviour than human beings (Saris, 1989) the participants agreed that this type of question can not be banned completely. It remains necessary in many studies and therefore should be one of the questions studied.

A special contribution in this respect was made by Melich of the EC committee, in charge of the Eurobarometer research. She was asked to draw our attention to the research her committee was doing in all EC countries. There are 11 questions which are asked in every wave of the Eurobarometer. Time trends can be made from 1973. Although these questions are asked repeatedly, it does not mean that they are very well formulated or that they have no methodological problems. She mentioned the following problems:

- The words in the questions and the labels of the scales mean different things for respondents of different countries.
  Question: Is it possible to use procedures which are less language dependent?
- The topics are not very familiar to the respondents and therefore need a lot of introduction.
  Question: How can this information be optimized?
- Sometimes errors occur in the formulation of questions or in the printing.
  Question: What is the effect of such errors on the responses?
- Recently the 4-point scales were substituted by 10-point scales.
  Question: How do the previous results compare with those obtained with the new scale?
- A general question is, of course, what would be the optimal scale for these questions?
- For some topics it is necessary to know whether the question applies to the respondent or not (for example possession of a tv).
  Question: Is a filter the best way or not and how does the difference in quality compare with the difference in price?

These were all seen as interesting topics for further consideration. It was clear to the participants that these questions of the EC would be worth studying. It was also clear that for at least some of the questions the answers could well be given by the kind of research we are planning to do. Further discussion between our research group and the Eurobarometer committee was decided upon.

Many more topics were of course mentioned in the discussion but they cannot all be mentioned. All participants filled in a form to provide an idea of the kind of variables they were especially interested in. The form provided a short list of topics suggested by Saris in his contribution. Analysis of the answers showed that for each of the variables at least one country had mentioned a special interest. Furthermore, we found that most participants are interested in new scaling procedures. This also emerged from the discussions we had during the meeting.

Given that none of the variables mentioned can be omitted because of lack of interest, it is clear that the research design of the meta-study is very important. There are very many factors which have to be varied and each of these has many possible values.

Choice of topics

In general, it was decided that the topics could be anything that suited our purpose. Nevertheless, participants have asked that special attention be given to specific topics which are of concern to them because of ongoing research. In this respect we can mention:

1. A subsidized project in Sweden concerning scaling mass communication variables in mail interviews.
2. An Italian project concerned with Post Materialism and other political issues.
3. The Danish group mentioned the European election study.
4. A Belgian project concerning attitudes towards African migrants.
5. The Spanish research group suggested that marketing and product evaluation questions be given special attention.

The discussion of the possible study of the Eurobarometer questions led to the idea that items which could be of more general interest than to only our "small" research group should be included. Therefore a suggestion was made to do a pilot study in which the topic "life satisfaction" would be chosen as the substantive carrier of the study. All participants agreed upon this choice.

CHAPTER 12

BASIC DECISIONS CONCERNING THE CHOICE OF VARIABLES

In order to perform the complex research project which we are planning, a number of decisions concerning the research process have to be made. These concern the following topics:
- quality indicators for measurement instruments
- the variables to be included
- the topics to be chosen
- the collection of the data
- the analysis of the raw data
- the construction of the data files
- the exchange of information

The quality criteria of the measurement instruments have been discussed in part 1. We shall also not discuss the data analysis here, but will concentrate on the choice of the variables which characterize the measurement instruments in the study and other points relevant to this study. These points are discussed in sequence.

VARIABLES TO BE INCLUDED IN THE STUDY

Many aspects of a survey design can be varied and consequently affect the quality of the obtained data. Andrews grouped these different aspects into four sets:
1. Question characteristics
2. Response scale characteristics
3. Research design characteristics
4. Question topic characteristics

Furthermore, he discussed characteristics of the respondents, the interviewers and the interview process which might have affected the quality of his data. This ordering is followed here but further characteristics drawn from other sources, such as Schuman and Presser (1981), Molenaar (1986), Billiet et al. (1984) and others, are also included.

CHARACTERISTICS OF SURVEY DESIGNS

There are four sets of characteristics which will be distinguished and they will be discussed in sequence. Where characteristics are nested in other characteristics, this will be indicated as it is important for the research design and the analysis.

Formulation of the question

Andrews suggested the following characteristics:

1. comparative- absolute judgment
2. length of question combined with length of introduction

On the basis of the studies of Schuman and Presser (1981), the following characteristics are added:

3. use of a statement or a question

3.1 questions can be unbalanced or balanced

3.2 questions can also contain only one or two arguments (for and against)

4. extremeness of the formulation

The possibilities given in 3.1 and 3.2 will not hold for statements. These two characteristics are nested in the question formulation.

Characteristics of the response scale

The number of characteristics of the response scale has been increased considerably:

1. Category scales or continuous scales can be used
   For category scales the following distinctions can be made:
   - open or closed

   If closed scales are used we have the possibilities:
   - labeled and partially unlabeled scales
   - variation in the number of categories

   For continuous scales there is a choice between:
   - lines and numbers
   - the number of points on the scales

Beside these nested characteristics there are also some general response scale characteristics which can be applied under all circumstances:

2. type of scale (agree-disagree/ for-against/ importance/ satisfaction)
3. treatment of don't know (included, filter, post hoc selection)
4. use of a scale midpoint
5. use of scales from negative to positive, zero to positive or zero to negative
6. use of a fixed connection between subjective and response scales

Research design

We suggest the following distinctions:

1. data collection method: face-to-face, mail, telephone
   For face to face interviewing we can distinguish:
   - self completion or not

while mail interviewing is automatically self completion and telephone interviewing is by necessity not self completion
2. use of computer assisted procedures or not
In principal, this distinction cannot be made for mail interviewing but if we consider tele-interviewing as a computer-assisted mail interview this is not a serious error.
3. the data can be collected from a fresh sample or within a panel design
4. the question can be an item in a battery or not
5. the question can vary with respect to position in the questionnaire

Topic of the question

For this set, Andrews suggested the following characteristics:
1. the social desirability related to the topic
2. the familiarity of the respondents with the topic
3. whether questions concern the past, present of the future
4. the saliency of the topic
In agreement with the approach of Molenaar, we would like to add at least a rough categorization of the topics themselves:
5. the content of the question

CHARACTERISTICS OF RESPONDENTS

As has been mentioned above, the characteristics of the respondents can also affect the quality of the instruments. Andrews has therefore introduced these characteristics in his analysis, but I recommend that we use this information slightly differently. Andrews analyzed many different subsamples, which could be overlapping. My suggestion is that this be avoided because of the dependencies which will be created between the "cases" in this analysis.

Nevertheless, we must take these characteristics of the respondents into account because the samples in the different countries may differ with respect to these characteristics. Also, subsamples selected because of lack of knowledge or for other reasons may be very different with respect to these characteristics. These characteristics could then explain the differences in quality of the instruments in the different countries and must therefore be controlled in the analysis.

I suggest the following:

Background characteristics
1. education
2. age

3. race
4. language
5. gender

Interest in and knowledge of the topic

1. a rating of interest by the interviewer and/or respondent
2. lack of knowledge

CHARACTERISTICS OF THE INTERVIEWERS

In addition to differences in the composition of the respondents, differences in the characteristics of the interviewers may affect the studies. We will make a distinction between background characteristics and experience.

Background characteristics

1. education
2. age
3. race
4. gender
5. language

Experience

1. number of interviews done in general
2. number of interviews within the project
3. interest in the topic

Characteristics of the interview process

The interview process may also differ from country to country and must therefore be controlled. In order to do so we have to evaluate the following aspects:

1. the number of requests for clarifications
2. the number of times needed to get in contact with a respondent
3. the concern for social desirability of the respondent
4. the amount of prior experience with interviews

THE TOPICS TO BE CHOSEN

With respect to the topics which can be chosen for this study, we should restrict ourselves to subjective variables such as:

- evaluations (good - bad)
- agreement with opinions (agree - disagree)
- agreement with actions (for - against)
- preferences
- evaluation of importance

- satisfaction
- frequencies

These kinds of questions can be asked for many different topics. I recommend that we choose topics which are of considerable importance in the European community or otherwise on an international level. Choosing such topics could facilitate obtaining grants for this study, especially if we use the same topics in different countries to allow a comparison of the results across countries. Another strategy would be to relate this research to ongoing research in each country. The cost of the research will then be lower because questions which have already been asked have also been paid for.

Both procedures guarantee that we will not collect data for variables which are only of academic interest. It does not mean that we can realize a random sample of questions, which would be difficult. Perhaps the best we can aim for is to collect information for variables which will be seen by others as relevant.

I would like to suggest the following topics:

1. Questions which have been asked frequently in the Eurobarometers:
   - evaluation of efforts to unify Europe
   - evaluation of membership of EC
   - importance of the European Parliament for the community
     etc.
2. Developments with respect to East-West relationships:
   - evaluation of the efforts towards disarmament
   - evaluation of efforts towards better international relationships
     etc.
3. Satisfaction with aspects of social life (social indicators)
   - satisfaction with housing
   - satisfaction with health
   - satisfaction with work
   - satisfaction with family life,
     etc.
4. Questions with respect to the economic situation (Katona items)
   - personal situation
   - country as a whole
   - future developments
   - confidence in the future,
     etc.

5. Evaluation of new technologies
   - nuclear energy
   - genetic engineering,
   etc.
6. Influence of population on government
   - influence of the population
   - preferred influence
   - evaluation of different possibilities
   etc.

Many more topics could be formulated along these lines. At least some topics should be chosen which will be asked in all participating countries in the same way. This will give us a base line for comparing the results obtained for other topics.

REFERENCES

Andrews, F.M. (1984). Construct validity and error components of survey measures: A structural modeling approach. *Public Opinion Quarterly, 48,* 409-442.

Belson, W.A. (1981). *The design and understanding of survey questions.* Farnborough: Gower.

Billiet, J., Loosveldt, G., & Waterplas, L. (1984). *Het survey-interview onderzocht: Effecten van het ontwerp en gebruik van vragenlijsten op de kwaliteit van de antwoorden.* Leuven: Sociologisch Instituut.

Molenaar, N.J. (1986). *Formuleringseffecten in survey-interviews (Wording effects in survey-interviews).* Amsterdam: VU-Uitgeverij.

Saris, W.E. (1988). *Variation in response functions: a source of measurement error in attitude research.* Amsterdam: Sociometric Research Foundation.

Saris, W.E. (1989). A technological revolution in data collection. *Quality and Quantity, 23,* 333-349.

Schuman, H., & Presser, S. (1981). *Questions and answers in attitude surveys.* London: Academic press.

Sudman, S., & Bradburn, N.L. (1974). *Response effects in surveys.* Chicago: Aldine.

Design of the international research project

From the discussions taking place during the conference in Amsterdam several conclusions have been drawn with respect to the design of the study, topics which require further study, the time planning and organization of the study, the participants. About each of these topics we will make a final remark.

THE DESIGN OF THE STUDY

It has been decided that for this study each country will do MTMM studies with at least three topics and three methods. The model which will be used is specified in detail in chapter 5. It will contain restrictions on the method effects and preferably no correlations between the method factors.

The studies will be done on preferably national samples, but local samples are not excluded. The analysis should be based on samples of minimally 500 cases. Sufficient time should be left between the repeated questions so that it is likely that the previous answers have been forgotten. The topics and methods are not constrained yet. Later we will try to stimulate co-ordination in order to approach an experimental design as closely as possible. The analysis will be done using the ML estimation procedure available in programs like LISREL and EQS.

Each country has the rights to publish the results of its studies along its own channels but the results of the studies will also be provided to the co-ordination centre in Amsterdam for the meta-analysis. This information should include data with respect to characteristics of the data collection procedure, the formulation of the questions, the characteristics of the sample and the interviewers. The co-ordination centre in Amsterdam will do the meta analysis if sufficient data is available and the groups publications will be prepared.

TIME PLANNING

The definite project will start in 1990. The year 1989 will be used for an international pilot study: every country will do at least one study trying out the MTMM design. The topic for the pilot study will be life satisfaction, especially four "aspects":
1. satisfaction with life in general
2. satisfaction with housing
3. satisfaction with contact with others
4. satisfaction with the financial situation of the household.

The methods varied in the pilot study will be:
- 5 points category scale
- 10 points category scale
- magnitude estimation with three reference points

From the beginning of 1990 till the end of 1992 we expect that each participant will do at least 10 different studies according to the above specified design and each participant will send his results to Amsterdam for further analysis. In 1993 the co-ordination centre in Amsterdam will perform the meta analysis on the existing data. The results will be discussed in a meeting in 1993 with the participants and afterwards be reported in a publication of the group. This does not mean that we think that the final word will be spoken about data quality at that moment as we will make clear in one of the next sections.

THE ORGANIZATION AND PARTICIPANTS

So far 20 different research groups from 14 different countries have agreed to cooperate in this project. The co-ordination of the project is in the hands of the Dutch research group. Below we give the names of the different participants and their research institution starting with the Dutch co-ordination centre. The work of the research group is co-ordinated by the distribution of a newsletter.

co-ordination centre:
W.E. Saris, and A. Scherpenzeel
University of Amsterdam, Methodology Department
Oudezijds Achterburgwal 237, 1012 DL Amsterdam, the Netherlands

other participants:
F. Aarebrot
University of Bergen, Institutt for Sammenlignende Politik
Christies gt 17-19, Bergen, Norway

F.M. Andrews
University of Michigan, Survey Research Center
Box 1248, Ann Arbor MI 48106, USA

V. Andreyenkov
USSR Academy of Sciences, Survey Methodology Department, Institute for Sociological Research
Krzhyzhanovskogo str. 24/35 # 5, Moscow 117259, USSR

J.M. Batista-Foguet
E.S.A.D.E.
Av. da. Pedralbes 60-62, 08034 Barcelona, Spain

H. Berger
The Academy of Science, Institute of Sociology and Social Policy
Otto Nusatzke strasse 10/11, Berlin, DDR-108

J.B. Billiet, and H. de Wit
University of Leuven, Department of Sociology
E.van Even straat 2C, B.3000 Leuven, Belgium

R. Cayrol, L. Blondiaux, and S. Duchesne
Fondation National des Sciences Politiques
27 Rue St.-Guillaume, 75007 Paris, France

A. Ferligoj
E.Kardels University, Faculty of Sociology, Political Science and Journalism
Kardelseva pl.5, 61000 Ljubljana, Yugoslavia

E. Gehmacher
IFES
Rainergasse 38, 1050 Wien, Austria

A. Kolstad
Hogskolesenteret I Nordland
Postboks 6003, 8016 Morkved, Norway

R. Költringer
University of Wien, Sociology Department
Alserstrasse 33, A-1080 Wien, Austria

D. Krebs
ZUMA
Postbox 125969, D6800 Mannheim 1, West Germany

C. O'Muircheartaigh, and G. Gaskell
London School of Economics, Center for Survey Teaching and Research
Houghton street WC2A 2AE, London, United Kingdom

I. Nicolacopoulos, K. Tsoukalas, and C. Michalopoulos
National Center of Social Research EKKE
1 Sophocleous st., 105 59 Athens, Greece

K. Reif, and A. Melich
Commission of the EC, "Surveys, Research, Analyses"
Berl.2/103, Rue de la Loi 200, 1049 Brussels, Belgium

K. Roe, and A. Ohlson
University of Gothenburg, Unit of Mass Communication
Box 5048, S-40221 Gothenburg, Sweden

H. Schadee
Unversità Degli Studi di Trento, Dipartemento di Politica Sociale
Via Verdi 26, I-38100 Trento, Italia

P. Schmitt
Justus-Liebig-Universität, Institut für Politikwissenschaft
Karl-Glöckner Str. 21, Haus E, 6300 Giesen, West Germany

O. Transgaard, and K. Siune
University of Arhus, Institute of Political Science
DK-8000 Arhus C, Denmark

Invitation to contribute to this research

Although we are very happy that already so many research groups have agreed to cooperate in this research project it does not mean that other groups can not participate. We are certainly interested in other participants if they are willing to work according to the procedures agreed upon.

We do not think that the final answer will be given in these three years on all questions concerning the quality of measurement instruments for all the possible variations in design of data collection procedures. Inventions in this field will always go on, and we think that this type of study should be updated occasionally. Therefore we will develop a database with information for meta-analysis which can easily be updated with results of new studies. Also all data analysis programs will be stored so that updating of the meta-analysis can be done with little effort whenever there are enough new results available.

This last section should stimulate any research group which is doing MTMM research to provide their results to our database for further analysis in a later stage. If research groups are interested in a more close connection with our research group or even to cooperate they can contact the co-ordination centre for further information.

Literature references with respect to the MTMM aproach

APPLICATIONS

Aaker, D.A., Bagozzi, R.P., Carman, J.M., & macLachlan, J.M. (1980). On using response latency to measure preference. *Journal of Marketing Research, 17,* 237-244.

Aneshensel, C.S., & Yokopenic, P.A. (1985). Tests for the comparability of a causal model of depression under 2 conditions of interviewing. *Journal of Personality and Social Psychology, 49,* 1337-1348.

Arora, R. (1982). Validation of an S-O-R model for situation, enduring, and response components of involvement. *Journal of Marketing Research, 19,* 505-516.

Atkinson, T. (1982). The stability and validity of quality of life measures. *Social Indicators Research, 10,* 113-132.

Bachman, L.F., & Almer, A.S. (1982). The construct-validation of some components of communicative proficiency. *Tesol Quarterly, 16,* 449-463.

Bachman, L.F., & Palmer, A.S. (1981). The construct-validation of the FSI oral interview. *Language Learning, 31,* 67-86.

Bagozzi, R.P. (1978). Construct-validity of affective, behavioral, and cognitive components of attitude by analysis of covariance-structures. *Multivariate Behavioral Research, 13,* 9-31.

Bagozzi, R.P., & Vanloo, M. (1987). Individual and couple tastes for children: Theoretical, methodological, and empirical issues. *Journal of Economic Psychology, 8,* 191-214.

Bagozzi, R.P., Tybout, A.M., Craig, C.S., & Sternthal, B. (1979). Construct-validity of the tripartite classification of attitudes. *Journal of Marketing Research, 16,* 88-95.

Breckler, S.J. (1984). Empirical validation of affect, behavior, and cognition as distinct components of attitude. *Journal of Personality and Social Psychology, 47,* 1191-1205.

Bryant, K.J. (1984). Methodological convergence as an issue within environmental cognition research. *Journal of Environmental Psychology, 4,* 43-60.

Burnkrant, R.E., & Page, T.J. (1982). An examination of the convergent, discriminant, and predictive-validity of Fishbein behavioral intention model. *Journal of Marketing Research, 19,* 550-561.

Byrne, B.M. (1986). Self-concept academic-achievement relations: An investigation of dimensionality, stability, and causality. *Canadian Journal of Behavioural Science - Revue Canadienne des Sciences du Comportement, 18,* 173-186.

Childers, T.L., Houston, M.J., & Heckler, S.E. (1985). Measurement of individual-differences in visual versus verbal information-processing. *Journal of Consumer Research, 12,* 125-134.

Crandall, R. (1976). Validation of self-report measures using ratings by others. *Sociological Methods & Research, 4,* 380-400.

Cummings, K.M., Jette, A.M., & Rosenstock, I.M. (1978). Construct-validation of the health belief model. *Health Education Monographs, 6,* 394-405.

Dean, A., & Lin, N. (1977). Stress-buffering role of social support: Problems and prospects for systematic investigation. *Journal of Nervous and Mental Disease, 165,* 403-417.

Dess, G.G., & Beard, D.W. (1984). Dimensions of Organizational Task Environments. *Administrative Science Quarterly, 29,* 52-73.

Edwards, J.E., Rode, L.G., & Ayman, R. (1989). The construct validity of scales from 4 leadership questionnaires. *Journal of General Psychology, 116,* 171-181.

Farh, J.L., Hoffman, R.C., & Hegarty, W.H. (1984). Assessing environmental scanning at the subunit level: A multitrait-multimethod analysis. *Decision Sciences, 15,* 197-220.

Fergusson, D.M., & Horwood, L.J. (1987). The trait and method components of ratings of conduct disorder: 1. Maternal and teacher evaluations of conduct disorder in young-children. *Journal of Child Psychology and Psychiatry and Allied Disciplines, 28,* 249-260.

Fergusson, D.M., & Horwood, L.J. (1989). Estimation of method and trait variance in ratings of conduct disorder. *Journal of Child Psychology and Psychiatry and Allied Disciplines, 30,* 365-378.

Fleming, J.S., & Courtney, B.E. (1984). The dimensionality of self-esteem: 2. Hierarchical facet model forrevised measurement scales. *Journal of Personality and Social Psychology, 46,* 404-421.

Fouad, N.A., Cudeck, R., & Hansen, J.I.C. (1984). Convergent validity of the Spanish and English forms of the strong-Campbell interest inventory for bilingual hispanic high-school-students. *Journal of Counseling Psychology, 31,* 339-348.

Francis, D.J., Fletcher, J.M., & Rourke, B.P. (1988). Discriminant validity of lateral sensorimotor tests in children. *Journal of Clinical and Experimental Neuropsychology, 10,* 779-799.

Freedman, R.D., & Stumpf, S.A. (1978). Student-evaluations of courses and faculty based on a perceived learning criterion: Scale-construction, validation, and comparison of results. *Applied Psychological Measurement, 2,* 189-202.

Friedman, C.P., Trier, W.C., & Forsythe, G.B. (1987). Confirmatory factor-analysis of house office candidate appraisals. *Evaluation & The Health Professions, 10,* 106-119.

Gabbard, C.E., Howard, G.S., & Dunfee, E.J. (1986). Reliability, sensitivity to measuring change, and construct-validity of a measure of counselor adaptability. *Journal of Counseling Psychology, 33,* 377-386.

Gabbard, C.E., Howard, G.S., & Tageson, C. (1986). Assessing locus of control with religious populations. *Journal of Research in Personality, 20,* 292-308.

Gaski, J.F. (1987). Construct-validity of measures of college-teaching effectiveness: Comment. *Journal of Educational Psychology, 79,* 326-330.

Graham, J.W., Rohrbach, L.A., Hansen, W.B., Flay, B.R., & Johnson, C.A. (1989). Convergent and discriminant validity for assessment of skill in resisting a role play alcohol offer. *Behavioral Assessment, 11,* 353-379.

Green, S.E., & Mosher, D.L. (1985). A causal model of sexual arousal to erotic fantasies. *Journal of Sex Research, 21,* 1-23.

Hattie, J. (1979). Confirmatory factor-analysis of the pat: Reading-comprehension, reading vocabulary, and listening comprehension tests. *New Zealand Journal of Educational Studies, 14,* 172-188.

Hertzog, C., Hultsch, D.F., & Dixon, R.A. (1989). Evidence for the convergent validity of 2 self-report metamemory questionnaires. *Developmental Psychology, 25,* 687-700.

Holzbach, R.L. (1978). Rater bias in performance ratings: Superior, self-ratings and peer ratings. *Journal of Applied Psychology, 63,* 579-588.

Houts, A.C., Cook, T.D., & Shadish, W.R. (1986). The person-situation debate: a critical multiplist perspective. *Journal of Personality, 54,* 52-105.

Howard, G.S., Conway, C.G., & Maxwell, S.E. (1985). Construct-validity of measures of college-teaching effectiveness. *Journal of Educational Psychology, 77*, 187-196.

Jaccard, J., & Sheng, D. (1984). A comparison of 6 methods for assessing the importance of perceived consequences in behavioral decisions: Applications from attitude research. *Journal of Experimental Ssocial Psychology, 20*, 1-28.

Jackson, D.N., & Paunonen, S.V. (1980). Personality structure and assessment. *Annual Review of Psychology, 31*, 503-551.

John, G., & Martin, J. (1984). Effects of organizational-structure of marketing planning on credibility and utilization of plan output. *Journal of Marketing Research, 21*, 170-183.

John, G., & Reve, T. (1982). The reliability and validity of key informant data from dyadic relationships in marketing channels. *Journal of Marketing Research, 19*, 517-524.

Judd, C.M., Jessor, R., & Donovan, J.E. (1986). Structural equation models and personality-research. *Journal of Personality, 54*, 149-198.

Lastovicka, J.L. (1982). On the validation of lifestyle traits: A review and illustration. *Journal of Marketing Research, 19*, 126-138.

Lastovicka, J.L. (1983). Convergent and discriminant validity of television commercial rating-scales. *Journal of Advertising, 12*, 14.

Leinhardt, G. (1980). Modeling and measuring educational treatment in evaluation. *Review of Educational Research, 50*, 393-420.

Marsh, H.W., & Butler, S. (1984). Evaluating reading diagnostic-tests: An application of confirmatory factor-analysis to multitrait-multimethod data. *Applied Psychological Measurement, 8*, 307-320.

Marsh, H.W., & Hocevar, D. (1984). The factorial invariance of student-evaluations of college-teaching. *American Educational Research Journal, 21*, 341-366.

Marsh, H.W., Antill, J.K., & Cunningham, J.D. (1987). Masculinity, femininity, and androgyny: Relations to self-esteem and social desirability. *Journal of Personality, 55*, 661-685.

Marsh, H.W., Barnes, J., & Hocevar, D. (1985). Self other agreement on multidimensional self-concept ratings: Factor-analysis and multitrait-multimethod analysis. *Journal of Personality and Social Psychology, 49*, 1360-1377.

Marsh, H.W., & O'Neill, R. (1984). Self description questionnaire: 3. The construct-validity of multidimensional self-concept ratings by late adolescents. *Journal of Educational Measurement, 21,* 153-174.

Menezes, D., & Elbert, N.F. (1979). Alternative semantic scaling formats for measuring store image: Evaluation. *Journal of Marketing Research, 16,* 80-87.

macNab D., & Fitzsimmons, G.W. (1987). A multitrait multimethod study of work-related needs, values, and preferences. *Journal of Vocational Behavior, 30,* 1-15.

Neibecker, B. (1984). The validity of computer-controlled magnitude scaling to measure emotional impact of stimuli. *Journal of Marketing Research, 21,* 325-331.

Newcomb, M.D., & Chou, C.P. (1989). Social support among young adults: Latent-variable models of quantity and satisfaction within 6 life areas. *Multivariate Behavioral Research, 24,* 233-256.

Olweus, D. (1980). Familial and temperamental determinants of aggressive-behavior in adolescent boys: A causal-analysis. *Developmental Psychology, 16,* 644-660.

Palmer, J., macLeod, C.M., Hunt, E., & Davidson, J.E. (1985). Information-processing correlates of reading. *Journal of Memory and Language, 24,* 59-88.

Quellmalz, E.S., Capell F.J., & Chou, C.P. (1982). Effects of discourse and response-mode on the measurement of writing competence. *Journal of Educational Measurement, 19,* 241-258.

Rau, W.C. (1980). The tacit conventions of the modernity school: An analysis of key assumptions. American Sociological Review, 45, 244-260.

Reuman, D.A., Alwin, D.F., & Veroff, J. (1984). Assessing the validity of the achievement motive in the presence of random measurement error. *Journal of Personality and Social Psychology, 47,* 1347-1362.

Ruekert, R.W., & Churchill, G.A. (1984). Reliability and validity of alternative measures of channel member satisfaction. *Journal of Marketing Research, 21,* 226-233.

Sawyer, D.O. (1980). Normative conflict and physician use: A latent structure approach. *Journal of Health and Social Behavior, 21,* 156-169.

Schibeci, R.A. (1984). Students, teachers, and the assessment of attitudes to school. *Australian Journal of Education, 28,* 17-24.

Silk, A.J., & Kalwani, M.U. (1982). Measuring influence in organizational purchase decisions. *Journal of Marketing Research, 19,* 165-181.

Slama, M.E., & Tashchian, A. (1985). Selected socioeconomic and demographic characteristics associated with purchasing involvement. *Journal of Marketing, 49,* 72-82.

Stacy, A.W., Widaman, K.F., Hays, R., & Dimatteo, M.R. (1985). Validity of self-reports of alcohol and other drug-use: A multitrait-multimethod assessment. *Journal of Personality and Social Psychology, 49,* 219-232.

Tate, U.S. (1984). Convergent and discriminant validity of measures of job, leisure, dyadic, and general life satisfaction by causal-modeling methodology. *Journal of Leisure Research, 16,* 250-254.

Turk, D.C., Rudy, T.E., & Salovey, P. (1985). The mcGill pain questionnaire reconsidered: confirming the factor structure and examining appropriate uses. *Pain, 21,* 385-397.

Tyler, T.R., & Rasinski, K. (1984). Comparing psychological images of the social perceiver: Role of perceived informativeness, memorability, and affect in mediating the impact of crime victimization. *Journal of Personality and Social Psychology, 46,* 308-329.

Upshur, J.A. (1983). Measurement of individual-differences and explanation in the language sciences. *Language Learning, 33,* 99-140.

Vance, R.J., macCallum, R.C., Hedge, J.W., & Coovert, M.D. (1988). Construct-validity of multiple job-performance measures using confirmatory factor-analysis. *Journal of Applied Psychology, 73,* 74-80.

Venkatraman, N., & Grant, J.H. (1986). Construct measurement in organizational strategy research: A critique and proposal. *Academy of Management Review, 11,* 71-87.

Venkatraman, N., & Ramanujam, V. (1986). Measurement of business performance in strategy research: a comparison of approaches. *Academy of Management Review, 11,* 801-814.

Venkatraman, N., & Ramanujam, V. (1987). Measurement of business economic-performance: An examination of method convergence. *Journal of Management, 13,* 109-122.

Ward, W.C. (1982). A comparison of free-response and multiple-choice forms of verbal aptitude-tests. *Applied Psychological Measurement, 6,* 1-11.

Weinrott, M.R., Jones, R.R., & Boler, G.R. (1981). Convergent and discriminant validity of 5 classroom observation systems: A secondary analysis. *Journal of Educational Psychology, 73*, 671-680.

Whitely, S.E. (1983). Construct-validity: Construct representation versus nomothetic span. *Psychological Bulletin, 93*, 179-197.

Williams, L.J., Cote, J.A., & Buckley, M.R. (1989). Lack of method variance in self-reported affect and perceptions at work: Reality or artifact. *Journal of Applied Psychology, 74*, 462-468.

Wingard, J.A. (1980). Life-span developmental-changes in mnemonic organization: A multimethod analysis. *International Journal of Behavioral Development, 3*, 467-487.

Zimmermann, E. (1980). Anomy, alienation, aggression: Plea for perspectives, effects in the political arena. *Beitrage zur Konfliktforschung, 10*, 113-129.

METHODOLOGICAL ARTICLES

Acito, F., Anderson, R.D., & Engledow, J.L. (1980). A simulation study of methods for hypothesis-testing in factor-analysis. *Journal of Consumer Research, 7*, 141-150.

Anderson, J.C. (1985). A measurement model to assess measure-specific factors in multiple-informant research. *Journal of Marketing Research, 22*, 86-92.

Anderson, J.C. (1987). An approach for confirmatory measurement and structural equation modeling of organizational properties. *Management Science, 33*, 525-541.

Anderson, J.C., & Gerbing, D.W. (1988). Structural equation modeling in practice: A review and recommended 2-step approach. *Psychological Bulletin, 103*, 411-423.

Andrews, F.M. (1984). Construct-validity and error-components of survey measures: A structural modeling approach. *Public Opinion Quarterly, 48*, 409-442.

Andrews, F.M., & Herzog, A.R. (1986). The quality of survey data as related to age of respondent. *Journal of the American Statistical Association, 81*, 403-410.

Avison, W.R. (1978). Auxiliary theory and multitrait-multimethod validation: A review of 2 approaches. *Applied Psychological Measurement, 2*, 433-449.

Bacher, F. (1986). Theoretical and observed variables: The multitrait-multimethod approach. *Revue de Psychologie Appliquee, 36*, 469-490.

Bagozzi, R.P. (1977). Structural equation models in experimental research. *Journal of Marketing Research, 14,* 209-226.

Bagozzi, R.P., & Phillips, L.W. (1980). Causal-models in marketing. *Journal of Marketing, 44,* 126-128.

Bailey, K.D. (1988). The conceptualization of validity: Current perspectives. *Social Science Research, 17,* 117-136.

Bentler, P.M., & Abramson, P.R. (1981). The science of sex-research: Some methodological considerations. *Archives of Sexual Behavior, 10,* 225-251.

Briggs, S.R., & Cheek, J.M. (1986). The role of factor-analysis in the development and evaluation of personality-scales. *Journal of Personality, 54,* 106-148.

Brinberg, D., & Hirschman, E.C. (1986). Multiple orientations for the conduct of marketing-research: An analysis of the academic practitioner distinction. *Journal of Marketing, 50,* 161-173.

Bycio, P., Alvares, K.M., & Hahn, J. (1987). Situational specificity in assessment-center ratings: A confirmatory factor-analysis. *Journal of Applied Psychology, 72,* 463-474.

Bynner, J. (1988). Factor-analysis and the construct indicator relationship. *Human Relations, 41,* 389-405.

Carroll, J.B. (1978). How shall we study individual-differences in cognitive-abilities: Methodological and theoretical perspectives. *Intelligence, 2,* 87-115.

Cole, D.A. (1987). Utility of confirmatory factor-analysis in test validation research. *Journal of Consulting and Clinical Psychology, 55,* 584-594.

Cole, D.A., & Maxwell, S.E. (1985). Multitrait-multimethod comparisons across populations: A confirmatory factor analytic approach. *Multivariate Behavioral Research, 20,* 389-417.

Comrey, A.L. (1988). Factor-analytic methods of scale development in personality and clinical-psychology. *Journal of Consulting and Clinical Psychology, 56,* 754-761.

Cudeck, R. (1982). Methods for estimating between-battery factors. *Multivariate Behavioral Research, 17,* 47-68.

Cudeck, R. (1988). Multiplicative models and mtmm matrices. *Journal of educational statistics, 13,* 131-147.

Cummings, K.M., Kirscht, J.P., Becker, M.H., & Levin, N.W. (1984). Construct-validity comparisons of 3 methods for measuring patient compliance. *Health Services Research, 19,* 103-116.

Cuttance, P.F. (1986). Towards a typology of information to aid reliability of response in social surveys. *Quality & Quantity, 20,* 27-52.

Davis, H.L., Douglas, S.P., & Silk, A.J. (1981). Measure unreliability: A hidden threat to cross-national marketing-research. *Journal of Marketing, 45,* 98-109.

Dillon, W.R., Madden, T.J., & Mulani, N. (1983). Scaling models for categorical variables: An application of latent structure models. *Journal of Consumer Research, 10,* 209-224.

Eiting, M.H., & Mellenbergh, G.J. (1980). Testing covariance-matrix hypotheses: An example from the measurement of musical abilities. *Multivariate Behavioral Research, 15,* 203-223.

Flamer, S. (1983). Assessment of the multitrait-multimethod matrix validity of Likert scales via confirmatory factor-analysis. *Multivariate Behavioral Research, 18,* 275-308.

Fornell, C., & Larcker, D.F. (1981). Evaluating structural equation models with unobservable variables and measurement error. *Journal of Marketing Research, 18,* 39-50.

Howard, J.A. (1987). The conceptualization and measurement of attributions. *Journal of Experimental Social Psychology, 23,* 32-58.

Hughes, M.A., Price, R.L., & Marrs, D.W. (1986). Linking theory construction and theory testing: Models with multiple indicators of latent-variables. *Academy of Management Review, 11,* 128-144.

Hui, C.H., & Triandis, H.C. (1985). Measurement in cross-cultural psychology: A review and comparison of strategies. *Journal of Cross-Cultural Psychology, 16,* 131-152.

Judd, C.M., & Krosnick, J.A. (1982). Attitude centrality, organization, and measurement. *Journal of Personality and Social Psychology, 42,* 436-447.

Kane, M.T. (1982). A sampling model for validity. *Applied Psychological Measurement, 6,* 125-160.

Kenny, D.A, & Berman, J.S. (1980). Statistical approaches to the correction of correlational bias. *Psychological Bulletin, 88,* 288-295.

Lee, S.Y. (1981). The multiplier method in constrained estimation of covariance structure models. *Journal of Statistical Computation and Simulation, 12,* 247-257.

Lehmann, D.R. (1988). An alternative procedure for assessing convergent and discriminant validity. *Applied Psychological Measurement, 12,* 411-423.

Lehrer, R., & Debernard, A. (1987). Language of learning and language of computing: The perceptual-language model. *Journal of Educational Psychology, 79,* 41-48.

Levin, J. (1988). Multiple group factor-analysis of multitrait-multimethod matrices. *Multivariate Behavioral Research, 23,* 469-479.

Lomax, R.G. (1986). The effect of measurement error in structural equation modeling. *Journal of Experimental Education, 54,* 157-162.

Marsh, H.W., & Hocevar, D. (1988). A new, more powerful approach to multitrait multimethod analyses: Application of 2nd-order confirmatory factor-analysis. *Journal of Applied Psychology, 73,* 107-117.

Marsh, H.W., & Hocevar, D. (1983). Confirmatory factor-analysis of multitrait-multimethod matrices. *Journal of Educational Measurement, 20,* 231-248.

Maruyama, G., & mcGarvey, B. (1980). Evaluating causal-models: An application of maximum-likelihood analysis of structural equations. *Psychological Bulletin, 87,* 502-512.

Mellenbergh, G.J., Kelderman, H., Stijlen, J.G., & Zondag, E. (1979). Linear-models for the analysis and construction of instruments in a facet design. *Psychological Bulletin, 86,* 766-776.

Ofir, C., Reddy, S.K., & Bechtel, G.G. (1987). Are semantic response scales equivalent. *Multivariate Behavioral Research, 22,* 21-38.

Phillips, L.W. (1981). Assessing measurement error in key informant reports: A methodological note on organizational analysis in marketing. *Journal of Marketing Research, 18,* 395-415.

Rezmovic, E.L., & Rezmovic, V. (1981). A confirmatory factor-analysis approach to construct-validation. *Educational and Psychological Measurement, 41,* 61-72.

Rodgers, W.L., Herzog, A.R., & Andrews, F.M. (1988). Interviewing older adults: Validity of self-reports of satisfaction. *Psychology and Aging, 3,* 264-272.

Sands, R., & Young, F.W. (1980). Component models for 3-way data: An alternating least-squares algorithm with optimal-scaling features. *Psychometrika, 45,* 39-67.

Schmitt, N. (1978). Path-analysis of multitrait-multimethod matrices. *Applied Psychological Measurement, 2,* 157-173.

Schmitt, N., & Stults, D.M. (1986). Methodology review: Analysis of multitrait-multimethod matrices. *Applied Psychological Measurement, 10,* 1-22.

Schriesheim, C.A., Solomon, E., & Kopelman, R.E. (1989). Grouped versus randomized format: An investigation of scale convergent and discriminant validity using LISREL confirmatory factor-analysis. *Applied Psychological Measurement, 13,* 19-32.

Shimp, T.A. (1983). A potpourri of consumer research methods. *Advances in Consumer Research, 10,* 127-129.

Vonsaldern, M. (1983). About the evaluation of multitrait-multimethod-matrices with LISREL. *Psychologische Beitrage, 25,* 255-279.

Widaman, K.F. (1985). Hierarchically nested covariance structure models for multitrait-multimethod data. *Applied Psychological Measurement, 9,* 1-26.